EAI/Springer Innovations in Communication and Computing

Editor's Note

The impact of information technologies is creating a new world yet not fully understood. The extent and speed of economic, life style and social changes already perceived in everyday life is hard to estimate without understanding the technological driving forces behind it. This series presents contributed volumes featuring the latest research and development in the various information engineering technologies that play a key role in this process.

The range of topics, focusing primarily on communications and computing engineering include, but are not limited to, wireless networks; mobile communication; design and learning; gaming; interaction; e-health and pervasive healthcare; energy management; smart grids; internet of things; cognitive radio networks; computation; cloud computing; ubiquitous connectivity, and in mode general smart living, smart cities, Internet of Things and more. The series publishes a combination of expanded papers selected from hosted and sponsored European Alliance for Innovation (EAI) conferences that present cutting edge, global research as well as provide new perspectives on traditional related engineering fields. This content, complemented with open calls for contribution of book titles and individual chapters, together maintain Springer's and EAI's high standards of academic excellence. The audience for the books consists of researchers, industry professionals, advanced level students as well as practitioners in related fields of activity include information and communication specialists, security experts, economists, urban planners, doctors, and in general representatives in all those walks of life affected ad contributing to the information revolution.

Indexing: This series is indexed in Scopus, Ei Compendex, and zbMATH

About EAI

EAI is a grassroots member organization initiated through cooperation between businesses, public, private and government organizations to address the global challenges of Europe's future competitiveness and link the European Research community with its counterparts around the globe. EAI reaches out to hundreds of thousands of individual subscribers on all continents and collaborates with an institutional member base including Fortune 500 companies, government organizations, and educational institutions, provide a free research and innovation platform.

Through its open free membership model EAI promotes a new research and innovation culture based on collaboration, connectivity and recognition of excellence by community.

More information about this series at https://link.springer.com/bookseries/15427

Mourade Azrour • Azeem Irshad
Rajasekhar Chaganti
Editors

IoT and Smart Devices for Sustainable Environment

Springer

EAI
RESEARCH MEETS INNOVATION

Editors
Mourade Azrour
Department of Computer Science
Faculty o fSciences & Techniques
Moulay Ismail University
Errachidia, Morocco

Azeem Irshad
Department Computer Science & Software Engineering
International Islamic University
Islama, Islamabad, Pakistan

Rajasekhar Chaganti
The University of Texas at San Antonio
San Antonio, TX, USA

ISSN 2522-8595 ISSN 2522-8609 (electronic)
EAI/Springer Innovations in Communication and Computing
ISBN 978-3-030-90082-3 ISBN 978-3-030-90083-0 (eBook)
https://doi.org/10.1007/978-3-030-90083-0

This Springer imprint is published by the registered company Springer Nature Switzerland AG
The registered company address is: Gewerbestrasse 11, 6330 Cham, Switzerland

Preface

The Internet of Things (IoT) has emerged as a prominent technology in recent years due to the integration of the physical objects and smart devices with the Internet. Thanks to the innovation in next-generation wireless technologies, Internet of Things is being widely implemented in nearly all domains of modern life, enabling individuals to work and live smarter. These domains encompass a wide range of industrial, strategic, and civil applications such as healthcare, smart homes, environment, transportation, and smart city. This advancement in technology opens up new avenues and opportunities to develop even more smarter devices including refrigerators, cookers, temperature controller, farming objects, medical devices, location trackers, and so on.

In environmental domain, Internet of Things has demonstrated its utility. Hence, in order to preserve natural resources and save the environment for the survival of human beings in the current and future generations, IoT can be used for monitoring and controlling numerous environmental objects. It is evident that the technology can be utilized to overcome the sustainable environment challenges. The recent advancements in smart farming are major example of how technology plays a significant role to preserve the environmental conditions for human life.

This book provides recent state-of-the-art researches related to the smart devices and IoT that are being applied in meeting the challenges to sustainable environment. The primary purpose of this book is to cover the state-of-the-art IoT applications, security in the IoT-based implementations, networking, machine learning, monitoring and controlling environment, smart metering, software and systems solution, and the applications of IoT ranging from smart home, smart health, and smart farming.

Organization of the Book

This book contains chapters authored by academic researchers, industry experts, and individual researchers in the fields of Internet of Things, environment, and computer sciences. The book demonstrates many illustrations along with solving the existing problems in Internet of Things–based applications, including security, and then covers the primary interest of the book, that is, application of the Internet of Things in smart farming focused on wastewater management and other topics in detail.

Acknowledgments

We would like to thank all the contributing authors for considering our special issue and the time, effort, and understanding during the preparation of this book.

Contents

Detection of Some Water Elements Based on IoT: Review Study 1
Fatimazahra Mousli, Jamal Mabrouki, Loubna Bouhachlaf, Mourade Azrour, and Souad El Hajjaji

Simulation of the Treatment Performance of a Purification Plant for a Dairy Effluent 19
Toufik Rachiq, Jamal Mabrouki, Souad El Hajjaji, and Sabir Rahal

Study of Piroxicam Removal from Wastewater by Artichoke Waste Using NemrodW® Software: Statistical Analysis 29
Nora Samghouli, Imane Bencheikh, Karima Azoulay, Fatima-Zahra Abahdou, Jamal Mabrouki, and Souad El Hajjaji

Theoretical Study of Thermoelectric Transport Properties of Dicalcium Silicide and Dicalcium Germanide Compounds 43
A. El Yousfi, H. Bouda, M. L. Ould Ne, A. G. El Hachimi, J. Mabrouki, A. El Kenz, and A. Benyoussef

Modeling and Design of Water Treatment Processes by Biosorption Method Using JMP® 11 Software 53
Karima Azoulay, Imane Bencheikh, Nora Samghouli, Jamal Mabrouki, Ahmed Moufti, and Souad El Hajjaji

Mathematical and Statistical Study for the Wastewater Adsorbent Regeneration Using the Central Composite Design 71
Imane Bencheikh, Karima Azoulay, Nora Samghouli, Jamal Mabrouki, Loubna Bouhachlaf, Ahmed Moufti, and Souad El Hajjaji

IoT and Reality Mining for Real-Time Disease Prediction 85
Hiba Asri

Smart System for Monitoring and Controlling of Agricultural Production by the IoT 103
Jamal Mabrouki, Karima Azoulay, Saloua Elfanssi, Loubna Bouhachlaf, Fatimazahra Mousli, Mourade Azrour, and Souad El Hajjaji

Simulation of Groundwater Quality: Case Study of the Limestone Chain of the Western Rif of Morocco 117
Ghizlane Fattah, Fouzia Ghrissi, Jamal Mabrouki, and Saloua Elfanssi

A Novel Anomaly Network Intrusion Detection System for Internet of Things Security 129
Azidine Guezzaz, Said Benkirane, and Mourade Azrour

Precision Agriculture: Assessing Water Status in Plants Using Unmanned Aerial Vehicle 139
Kaoutar Lkima, Francisco Pedrero Salcedo, Jamal Mabrouki, and Faissal Aziz

Smart Healthcare Using Evolutionary Algorithm 155
Deepti Saraswat and Manik Lal Das

Index 183

Detection of Some Water Elements Based on IoT: Review Study

Fatimazahra Mousli, Jamal Mabrouki, Loubna Bouhachlaf, Mourade Azrour, and Souad El Hajjaji

1 Introduction

The Internet of Things (IoT) has evolved with the development of wireless sensor network technologies, which makes communications between different sensors with the ability to capture, collect, transfer, and share data [1].

IoT has a great deal of ability to increase the quality of life of people by enhancement of the atmosphere through enhancing air quality, water quality monitoring, and emission mitigation. The deployment and construction of stations around the city and mounting sensors will do this. IoT expects significant advances in enhancing air and water quality as it provides the possibility of using state of the art measurement instruments. By using IoT, potential environmental developments can be predicted on the basis of space and time maps and natural hazards detected so that a lot of lives and large wealth can be saved. [2]. In the last century, the rise in human activities had a catastrophic environmental effect and has a significant impact on our well-being [3]. Water contamination is one of the most severe forms of environmental pollution, since any change in water quality has a direct effect on human health and the ecological balance. As a result, it is important to track and control toxins in real time to ensure environmental protection [4]. The lack of sanitation facilities and increased water use in businesses, especially in developed

F. Mousli (✉) · J. Mabrouki · L. Bouhachlaf · S. El Hajjaji
Laboratory of Spectroscopy, Molecular Modeling, Materials, Nanomaterial, Water and Environment, CERNE2D, Mohammed V University in Rabat, Faculty of Science, Agdal, Rabat, Morocco
e-mail: Fatimazahra.mousli@um5r.ac.ma

M. Azrour
Department of Computer Science, Faculty of Sciences & Techniques, Moulay Ismail University, Errachidia, Morocco

M. Azrour et al. (eds.), *IoT and Smart Devices for Sustainable Environment*, EAI/Springer Innovations in Communication and Computing,
https://doi.org/10.1007/978-3-030-90083-0_1

countries, lead to negative environmental impacts. Water quality is dependent on a series of criteria that are summarized into three major characteristics: physical, chemical, and biological parameters [5]. It includes details on the location and measurement of a degree of contamination.

Traditional methods of water quality monitoring have many disadvantages. First of all, it takes a long time and requires a lot of effort. Additionally, the expense of this approach is extremely high. Thirdly, the classical method done not offers real-time data, which may be helps for making significant and vital decisions [6]. As a result, having efficient surveillance systems built on the Internet of Things is crucial and necessary for ensuring environmental sustainability. This chapter presents an overview and description of the existing applications of water quality control systems based on wireless sensor networks and outlines different network and connectivity techniques, energy storage, and data processing techniques.

The rest of this chapter is organized as follows. In Sect. 2, we develop a general background for water parameters, and we describe the various network technologies and data processing techniques. In Sect. 3, we present the main features of the comparison of different developed systems. In Sect. 4, we present a detailed overview of the latest state-of-the-art monitoring of water quality by wireless sensor technology, which highlights important design choices for various implementations. Section 5 provides us with the conclusion and discussion of current open problems and potential priorities for study to improve existing wireless sensor network-based water quality monitoring systems.

2 Background

(a) *Water Parameters*

To check water quality, we will present the main measured parameters in most water quality monitoring systems despite the nature and type of applications:

- *pH sensor:* This is a measurement that determines whether a solution is acidic or basic (e.g., water). It measures between 0 and 60 °C and takes values between 0 and 14. When an acid dissolves in water, it releases the hydrogen ion (H+), while a base releases the hydroxyl ion (OH−). We used a pH meter to identify the pH value of the water being tested because the pH value is so important in assessing the acidity or basicity of water.
- *Turbidity sensor*: The relative clarity of the water is determined by this qualitative attribute. It refers to the presence of large, solid objects in water that have a negative impact on underwater life in rivers, lakes, and seas by obstructing the passage of sunlight to submerged aquatic plants. As a result, photosynthesis will cease, and dissolved oxygen levels will decrease.
- *Dissolved oxygen sensor*: Presents the free oxygen present in water. Because of the importance of it to all living things. The amount of oxygen in the water is

an important criterion for determining its quality. Its value should assist us in determining whether or not the water is of good quality.

- *Temperature sensor*: Water temperature is a physical characteristic describing the thermal energy of water, and it is an important parameter to consider when assessing water quality. The impacts of changing water temperatures on aquatic life are numerous. Furthermore, it has an impact on other variables.
- *Conductivity sensor*: A conductivity in homogenous content is equal to the conductance of a tube-shaped conductor made by the same material, divided by its surface, and repeated by its circumference.

(b) *Networking Technologies*

Networks help connect equipment to each other, such us the connection between sensors and other "objects." [7]

Each network has its strengths and weaknesses. However, each technology can be considered according to the following criteria:

- *Transmission mode:* The physical medium through which the transmission of information is carried out. This can be segmented into two categories:
  - Wireless communications: describes different radio communications such as: LAN (local area network), Bluetooth[8], Zigbee[9], IEEE 802.15.4 [10], Wi-Fi [11], WAN (wide area network), GSM/GPRS [12], 3G [13], 4G [14], LPWAN (low-power wide-area network), and LoRa [15].
  - Wired communications: they are mainly carried out over the three media (twisted pair cable, PLC, and optical fiber).

- Topology: It is necessary to distinguish the physical topology, which is the form that the network takes according to the nodes and their connections and the logical topology [16], which is the way in which the entities communicate such as the star, cellular, mesh, bus, etc. [17].
- *Range:* The transmission range of a signal, i.e., the maximum gap to interpret the signal of a receiver, can be short, medium, or large [18].
- *Bandwidth/capacity (throughput):* The larger the bandwidth, the higher the power output. This is due to the fact that there are several frequencies on which it is possible to transmit or receive several bits of information simultaneously [19].
- *Energy consumption:* In order to be able to assume the quality of service and to ensure the sending of more or less consequent volumes of DATA on a worldwide coverage, network requires energy [20]. Therefore, low-frequency technologies attest to a slightly longer lifespan of the "Devices" connected to their network, due to their low data transmission capacity [21].
- *Frequency:* In real time, once per hour or per day.

(c) *Data Processing*

The third essential component is data processing. In order to create value for the users of this data, it is absolutely necessary to store, archive, and backup it in databases and to structure it correctly. There are four main types of microcontrollers such as: ATMega328P, SAMD21, STM32F103C8T6, and STM32F407VET6.

3 Features of Comparison

The analysis of water quality by sensors has many advantages such as high sensitivity, good selectivity, speed, rapid response, etc. [22]. Different applications are developed due to the progress of IoT in water quality monitoring systems [23] such as:

- Monitoring fresh water sources like rivers, lakes, ponds, reservoirs or deep ground waters, etc. In order to monitor fresh water sources, different factors should be taken into consideration; first is the importance of identifying the nature of the freshwater, if it is static or moving and if its depth is in the surface or deep. Moreover, the sensing capabilities of nodes, type of communication, processing data, and network topology [24].
- Monitoring drinking water distribution system or water pipeline from the production plant to the final users, water travels through a large network of pipes of different sizes and diameters, which contamination is susceptible to increase and spread [25]. To prevent pipeline contamination, it is necessary to anticipate and manage systematically the protection of drinking water networks. Water quality monitoring using wireless sensor network in pipelines helps in protecting the health of potable water consumers and enhancing the economy using low-cost devices [26], but mostly helps in detecting pipe leakage, preventing incidents, and protecting the most valuable resource, which is potable water.
- Monitoring aquaculture systems. Water quality is at the heart of efficient fish farming. Fish rose in aquaculture live, breathe, and eat in the water, so they end up discharging waste. Water quality in ponds is a critical point in the production process and must be monitored in physical, chemical, and biological parameters. These must remain within acceptable limits for the proper development of organisms. Otherwise, the growing population could have poor growth, proliferation of pathogens with epidemics, and possible mortality and poor quality of the final product.

Therefore, it is necessary to implement a system to monitor water quality, taking into account the design of the farm and the origin of the water [27]. To ensure environmental sustainability, it is critical and essential to have effective monitoring systems based on the Internet of things. Environmental monitoring systems have been developed for monitoring water quality in a wide range of applications, which shows effective and high-performance results.

4 Results and Discussion

Actually, the development of new recent technologies and intelligent systems has improved the manner we organize our daily activities and actions. These smart systems have been developed based on the Internet of Things, which means a large network connecting various objects. These last ones have three main functions and roles. The first one is to sense and measure values from the area where they are placed. The second one is processing measured and captured information. While the last one is communicating the processed data to other linked devices.

There are various studies of water quality monitoring system that uses wireless sensor network technology in the Internet of Things in different fields. In this section, we will review these systems based on water application.

(a) *Natural Water Resources Monitoring Systems*

For *Natural water resources* in Table 1, *Yue and Ying* [28] measure the dissolved oxygen, pH, and turbidity; the transmission is done with the Ethernet; this system offers a sustainable advantage; it charges its battery continually using the solar power energy, in order to allow the system to operate for a long time. It is also based on clean energy. However, this system does not use any new technology, such as the microcontroller that processes data. Without considering security concerns, the author recommended remote access to the device over the Internet. This methodology presents vulnerabilities, which can influence measurement data's authenticity, honesty, and confidentiality after cyber security attacks. *Deqing et al.* [29] proposed an automated water quality monitoring and reporting device based on GSM which can capture pH, conductivity, dissolved oxygen, and turbidity values. The framework proposed is created on the basis of the data processing microcontroller and GSM module for the prediction of the data obtained and analyzed. Without considering security concerns, the author recommended remote access to the device over the Internet. This methodology presents vulnerabilities, which can influence measurement data's authenticity, honesty, and confidentiality after cyber security attacks. However, this system is expensive, and it cannot be afforded by common people. *Vijayakumar et al.* [30] proposed a temperature, pH, conductivity, dissolved oxygen, and turbidity measurable method. Raspberry is used to process data; the system transfers the data using WiFi. But the system has to be run under Linux kernel, so the user must write commands using keyboard every time to know the sensors reading. Without considering security concerns, the author recommended remote access to the device over the Internet, as well as the non-availability of a tool for data storage. *Aswale et al.* [31] proposed system measures of temperature and pH; the system uses Arduino Atmega328 microcontroller to process data. Even in this system, it can be installed in any geographical space with a few modifications. But only two criteria are measured, which are not adequate to note the positive or wrong nature of the water. Operating cost of this system is too expensive. All parameters to be tracked using a single node is unsafe as an anodic error allows all water quality data to be lost. The author recommends remote access

Table 1 Natural water resources monitoring systems

References	Measured parameters	Data processing	Data communication	Data storage	Energy management	Data security	Autonomy
Yue and Ying [28]	pH, turbidity	SunSPOT sensor	IEEE 802.15.4	None	Solar panel 13.5 V, 1.5 W; Battery 12 V	None	Cloudy for 100 h and sunny for 30 days
Mo Deqing et al. [29]	pH, conductivity, Dissolved O2, turbidity	Single chip/CD4051 switch	GSM/GPRS	None	Lithium and solar batteries combined	None	Cloudy for 100 hours and sunny for 30 days
Vijayakumar et al. [30]	T°, pH, conductivity, Dissolved O2, turbidity	Raspberry PI modèle B+ micropro-cesseur	USR WI-FI 232	None	Unspecified	None	Unspecified
Aswale et al. [31]	T°, pH	ARDUINO ATMega 328	Zigbee/GSM	None	Unspecified	None	Unspecified
Ranjbar and Abdalla [32]	T°, pH, turbidity, water level	Arduino ATMega 2560	GSM/GPRS	None	Battery 7-12 V Battery 5 V	None	Unspecified
Jamal Mabrouki et al. [33]	pH, temperature, conductivity, turbidity, Dissolved O2	Arduino uno ATmega328	ESP8266 Wi-Fi/HM-10 Bluetooth	Cloud	Unspecified	None	Unspecified
Sathish Pasika et al. [34]	pH, temperature, humidity, turbidity, water level	Arduino Mega	ESP8266 Wi-Fi	Cloud	Unspecified	None	Unspecified
Monira Mukta et al. [35]	pH, temperature, conductivity, turbidity,	Arduino Uno	USB	None	Unspecified	None	Unspecified
Vaishnavi V et al. [36]	pH, temperature, turbidity, flow	Arduino uno ATmega328	ESP8266 Wifi	None	Unspecified	None	Unspecified

to the device over the Internet without taking safety concerns into account, as well as the non-availability of a tool for data storage. *Ranjbar and Abdalla* [32] projected a system measuring pH, temperature, turbidity, and ultrasonic sensor. Measuring the water level, to process data the sensors are connected to a microcontroller type Arduino Mega 2560. The device has been tested with numerous impurities under varying conditions, water solutions, and at different times and can send water-measured values to final user through an SMS message. However, the system is not able to store the measurements *Jamal Mabrouki et al.* [33]. The author offers a minimal autonomous interface which shows water parameters on a small telephone screen, while allowing remote information to be transmitted to users. In addition, it can provide water quality estimates with high accuracy *Sathish Pasika et al.* [34]. The device proposed for real-time water quality monitoring is a cost-effective IoT approach. It is used to track the parameters pH, water turbidity, tank water level, temperature, and humidity of the environment. Arduino mega MCU unit aim boards integrate successfully with numerous sensors for building device measurement. An effective algorithm for monitoring water quality is built in real time, as well as an ESP8266 Wi-Fi module for data transmission. The analysis of many other factors such as electric conductivity, free residual chlorine, nitrates, and dissolved oxygen in water also needs to be carried out *Monira Mukta et al.* [35]. This work aims to track the quality of water samples by designing and evaluating the importance of these parameters using an effective machine learning method by designing a smart water quality monitoring unit integrated into the IoT platform, which detects four physical parameters: temperatures, pH, turbidity, and water conductivity. Various water samples are tested with the aid of Arduino sensors, and their various metrics are obtained. Fast forest binary classification demonstrates improved screening efficacy in prediction of water quality, validating the precision and efficiency of the system. This device can be deployed soon for real-time water surveillance solution with the update function of IoT technology for detecting chemical parameters of water *Vaishnavi V et al.* [36]. The proposed system monitors turbidity, pH, temperature, and flow of water. The data are processed via Arduino uno microcontroller. The data gathered and results of research are then made accessible by Wi-Fi module to the end user. Thus, more parameters need to be detected, and parameters increased by adding multiple sensors for the safest use.

(b) *Surface and River Water Quality Monitoring Systems*

For *surface and river water* in Table 2, *Amruta and Satish* [37] proposed the wireless sensor network for water quality surveillance system. A prototype of one node measuring pH, temperature, and dissolved oxygen solar cell-powered was created using ZigBee in order to send pH, turbidity, and level of oxygen sensor data to the base station. The results of the experiments have been used with a VB 6.0 GUI for analyzes of water quality data. Nevertheless, the prototype device had access by direct link to the base station to water quality data and gave no remote management to facilitate proactive reactions to water pollution. For the analysis of water quality data, a graphical interface was used, and graphic results were produced hourly using MATLAB. However, the sampling system provided access to water

Table 2 Surface and river water quality monitoring systems

References	Measured parameters	Data processing	Data communication	Data storage	Energy management	Data security	Autonomy
Amruta and Satih [37]	pH, Turbidity, oxygen level	LPC2148 MCU (Philips :NXP Semi-conductor)	ZigBee Module	None	Solar panel 13.5 V, 1.5 W ; Battery 12 V	None	Unspecified
Khetre and Hate [38]	T°, water level, turbidity, salinity	ARM-7 MCU	ZigBee	None	unspecified	None	Unspecified
Chung and Yoo [39]	T°, pH, conductivity, dissolved oxygen, turbidity, depth	ATmega 128	IEEE 802.15.4	None	Battery 12 V Solar panel	None	Unspecified
Samsudin et al. [40]	pH, turbidity	Arduino WeMos D1	WIFI ESP-12F	None	unspecified	None	Unspecified
Alexander T. et al. [41]	pH, temperature, dissolved oxygen	Arduino Mega 2560	Module GSM zigbee	My SQL; SD card	Solar Panel 10 W ; Battery 6V	None	Unspecified
Kofi Sarpong Adu-Manu et al. [42]	Temperature ;Conductivity Calcium; Nitrate; Fluoride ORP; Dissolved oxygen	Atmel ATMEGA 1281	4G (GSM)/ IEEE802.15.4 ZigBee	Cloud	Solar panel; rechargeable Battery	None	Unspecified
Paul D. et al [43]	Temperature; turbidity; pH; Dissolved oxygen	Arduino uno	GSM sim808	None	Solar panel; rechargeable Battery	None	Unspecified

quality data through a direct link to the base station and did not provide remote control to facilitate preventive water pollution responses. These systems have a short range of only a few decades, which reduces the system's spatial resolution. Usually, the range is broadened by using additional relay nodes, which generate additional routing costs. These strategies typically function in the frequency band 2.4 GHz, which is often blocked and susceptible to intrusion and attacks on defense. *Khetre and Hate* [38] developed a system for monitoring lake water quality by measuring temperature, salinity, water level, and turbidity. ARM 7 slave nodes are used to measure water quality parameters. ZigBee was designed to relay data to the Master Terminal using a two-slave prototype. The system has been checked and a 30.6 NTU reading obtained. The system has been evaluated. All other parameters on a live graph were recorded. Without considering security concerns, the author recommended remote access to the device over the Internet. This approach produces flaws that may impact calculation data authenticity, credibility, and secrecy during cyber security attacks. *Chung and Yoo* [39] developed a system that measures pH, temperature, conductivity, dissolved oxygen, turbidity, and depth of the river. The system uses Arduino Atmega128 microcontroller to process data. IEEE 802.15.4 is used to interconnect the collected data. Based on the experimental results, the author reported that the average water quality data were conveyed every 5 min and that the rate of data transmission damage was less than 1%. The consistency of communications provided by a flood system results in additional processing, information overloads that result in big energy expenses, memory costs, and inefficient bandwidth for wireless sensor network because their resources are limited *Samsudin et al.* [40]. Smart testing of a pH and turbidity measuring water quality detection device was added. Arduino processes the captured data and transmits it to the database through Wi-Fi. Overall, the device built provides rapid and simple pH and turbidity control to ensure that the water stays safe at all times. The system only covered the physical parameters of the water, so it must extend to the chemical parameters and the history of the readings is not available, as well as not taking into account the security issue of the data. *Alexander T. et al.* [41] proposed a low-cost system measuring the temperature, dissolved oxygen, and pH; Arduino Mega2560 is used to process data; the system transfers the data using Zigbee and GSM module. This low-cost system is appropriate to a wide range of coverage. The extension of the coverage area of this system covers big areas such as lakes and any form of water that needs continuous monitoring because of its importance. Consideration should also be provided to the usage of lithium-ion battery as a power supply and heavy metals as parameters of concern, as well as the consideration of data security issues for the expansion of the study. *Kofi Sarpong Adu-Manu et al.* [42] proposed a system detecting the conductivity, temperature, dissolved oxygen, and other chemical parameters. To process data, Atmel ATMega1281 is used; the system transfers the data using GSM and Zigbee module. The system provides an efficient and sustainable approach to monitoring freshwater sources; an improvement in the number of sensor nodes is to be planned taking into account sensor samples including turbidity and total dissolved solid sensors, as these are important water quality detection parameters, as well as taking into account the data security issue for the expansion of the study. *Paul D. et al.* [43] presented a comprehensive design and validation of an intelligent network of wireless sensors in order to measure a river's contamination levels. As the river case study showed high levels of pollution, several wireless sensor network nodes were implemented for obtaining information about the temperature, turbidity, pH, and dissolved oxygen. The device can also transmit information from the wireless sensor network node to the server via GSM. The data is available on-site. A large number of parameters and larger coverage should be taken into account.

(c) *Drinking Water Distribution Network Monitoring Systems*

For *drinking water distribution networks* in Table 3, *Faustine et al.* [44] developed a wireless sensor network solar energy system in the lake t o monitor water quality and measure the following: pH, temperature, conductivity, and dissolved oxygen. The Arduino core was used before it was sent to the gate via Zigbee to process measurement data. The portal gathered all of the data and distributed it to the GPRS program. Authors proved that the suggested system is functionally and practically applied in a specific environment from field test data. However, the framework does not provide for local data analysis, because any time a mobile network failure is encountered, it is offline *Lambrou et al.* [45]. A low-cost in-pipe sensor system was planned and built for measuring the quality of drinking water at customer sites. PIC32 and ARM/Linux microcontroller architectures were in use, respectively, for the system node and base station. Analysis shows that the device can detect high-impact microbiological (*Escherichia coli*) and chemical (*Arsenic*) pollutants at relatively low concentrations. Monitoring all parameters in a single node is therefore dangerous as the failure of all water quality data results from a nodal breakdown. Without contemplating security concerns, the machine provides remote system access through the Internet. This approach poses flaws that could impact the genuineness, credibility, and privacy of calculation information as a result of cyber security attacks. *Pradeepkumar et al.* [46] offered a low-cost system that can measure the temperature, pH, and turbidity; Arduino UNO is used to process data; the system transfers the data using GSM/GPRS. However, the systems measure only the water physical characteristics. The system measures physical parameters, which are insufficient for water quality monitoring, but the history of the readings is not available, as well as the lack of consideration of data security issues. *Simitha KM et al.* [47] proposed a system that measures the pH, turbidity temperature, and dissolved oxygen; Arduino ATmega328 is used to process data; the system transfers the data using WIFI and LoRa module. The system worked well as designed. The LoRa receiver efficiently received the values from the water quality sensors of the sending LoRa several kilometers away. The System is not capable of storing the measurements. *Marco Carminati et al.* [48] proposed a system that can measure the temperature, pH, conductivity, pressure, and water flow; to process data, Arduino ATMega328P is used; the system transfers the data using GSM/GPRS module. The results both in laboratories and on the ground with a three-node pilot network deployed in a real-world water delivery network and continuously operating for 2 months showed the robustness of the proposed device

Table 3 Drinking water distribution network monitoring systems

References	Measured parameters	Data processing	Data communication	Data storage	Energy management	Data security	Autonomy
Faustine et al. [44]	T°, pH, conductivity, Dissolved O2	Arduino Mega 2560	ZigBee/GPRS	None	Battery 3.7 V, 6AH; Solar panel 10 W	None	Unspecified
Lambrou et al. [45]	T°, pH, conductivity, Turbidity	PIC32 MCU /linux microcontroler	ZigBee	None	Unspecified	None	Unspecified
Pradeepkumar et al. [46]	T°, pH, turbidity	Arduino ATmega 328	GSM/GPRS	None	Battery 7–12 V	None	Unspecified
Simitha KM et al [47]	pH, temperature, turbidity, dissolved O2	ATMega 328	LoRa module RA-02 ; ESP32 Wi-Fi	None	Unspecified	None	Unspecified
Marco Carminati et al. [48]	pH, temperature, conductivity, pressure, flow	ATmega 328P	GSM/GPRS	Cloud	Turbine 5–12 V/Battery 5 V	None	Unspecified
Laiqa Binte et al. [49]	pH, temperature, dissolved solid, dissolved O2, turbidity.	Raspberry Pi3	GSM/LORA	Cloud	Solar panel/rechargeable battery	None	Unspecified

and the autonomy of energy through the effective solution of energy harvest for compact business turbines. Furthermore, the lack of costly components will cause the detection nodes to diffuse omnipresently, as well as the consideration of the data security issue for the expansion of the study. *Laiqa Binte et al.* [49] proposed a system that can collect the turbidity, temperature, pH, dissolved oxygen, and TDS. To process data, Raspberry Pi3 is used; the system transfers the data using GSM and LoRa module. The effects of the parameters are correct. The results are sent as an SMS on Android phones to upgrade the customer. People can also see the effects on the web page to become aware of the water quality. However, energy capabilities are easily reduced by the Raspberry Pi approach, as in addition to running the program code; a microprocessor needs to run a LINUX operating system. Cloud storage requires the use of encryption measures, as sensory data is open to future cyber threats, and this adds to increased computing, space, memory and energy costs *Aya Ayadi et al.* [50]. Different control strategies necessary for a water pipeline to ensure a high level of operation for end users are established as the primary focus in this work. In these current studies, the analysts discussed the various perspectives in which the problem was conceived. Furthermore, they give for these schemes a basic and concise guideline. They then proposed a comparative study performed in terms of various core characteristics that help define the progress and viability of various models.

(d) *Aquaculture Monitoring Systems*

For *aquaculture* in Table 4, *Wang et al.* [51] proposed a system that can send captured measures of temperature, pH, conductivity, and dissolved oxygen to database via Internet network using Zigbee and GPRS; however, this system does not use any new technology, such as the microcontroller that processes data. The scope of this technology is restricted to only a few decades, which reduces the system's spatial resolution. Typically, the range is expanded by the use of more relay nodes, which generate extra routing costs. The system is susceptible to attacks and interference. System is for use in aquaculture only. *Rao et al.* [52] proposed a device capable of measuring the potential for temperature, pH, conductivity, dissolved oxygen, and light oxidation. The captured values are then moved to a local computer to be stored. The processing data is based on Arduino Mega 2560; however, this system has to be linked directly to the computer through USB port, and it cannot be operated remotely. A computer has been programmed to receive Arduino data via USB. The system is vulnerable to interference and security attacks. The system is for use only in the aquaculture field. *Hongpin et al.* [53] proposed a device for tracking aquaculture water quality, analyzing temperature, pH, oxygen dissolving, and ammonia nitrogen for data processing, using an STM32F103 chip when passing the extracted value to storage and surveillance parties, which is responsible for GPRS and Zigbee. The system has been deployed in water aquaculture with a surface area of 2 hm^2, and its performance has been tested. The average loss rate in the network was 0.43%, meaning that data transmission was stable and reliable. The framework was associated with an aerator to perform programmed control of the disintegrated oxygen focus. The creator proposed distant admittance to the

Table 4 Aquaculture monitoring systems

References	Measured parameters	Data processing	Data communication	Data storage	Energy management	Data security	Autonomy
Wang et al. [51]	T°, pH, conductivity, dissolved oxygen	Simple electrical circuit	Zigbee/GPRS	Control center (PC); Internet	Unspecified	None	Unspecified
Rao et al. [52]	T°, pH, conductivity, dissolved oxygen, turbidity, salinity, redox potential, light	Arduino Mega 2560	USB	SQL database	Unspecified	None	Unspecified
Hongpin et al. [53]	T°, pH, dissolved oxygen, ammonium	STM32F103	Zigbee/GPRS	None	Solar panel 9 V/Lithium Battery 3,7 V	None	Unspecified
Parra et al. [54]	T°, conductivity, turbidity	Arduino Mega 2560	WI-FI ESP8266	Cloud	Unspecified	None	Unspecified

framework by means of the Internet without considering security issues, as well as the nonavailability of a tool for data storage. *Parra et al.* [54] developed a system collecting the temperature, conductivity, and turbidity; the captured data are stored in the cloud; the system uses Arduino Mega 256 to process the collected data and a module of Wi-Fi to transfer data, but this system measure only the water's physical characteristics. The system can monitor variations in water parameters, reservoir condition, and fish behavior during the nourishing process. System measures only the physical characteristics of the water. System is for use only in aquaculture.

5 Conclusion and Future Research

We offered a summary in the previous sections that described studies and implementations; we define the latest water quality monitoring frameworks, communication techniques, various network architectures, energy requirements, data processing processes, and water quality measurements.

With IoT and big data, analytics' exponential development as new innovations and their objective of facilitating multiple metropolitan centers in the sense of environmental sustainability and significant technological and technological issues must be explored and answered. There are mainly scientific, computational, and analytical problems. They are as follows:

- Communication and transmission of data
- Storage and processing data
- Security and privacy of data
- Sensor calibration
- Sensor fouling
- Energy management

Despite the fact that there have been a few progressions in water quality checking throughout the long term, there are a scope of open worries that need more examination to propel the utilization of remote sensor network-based water quality observing frameworks, as illustrated. In the water quality management process, data and network security are critical. Future applications can take into account issues such as malware attackers, technology breakdown, eavesdropping, and traffic monitoring. Furthermore, energy storage strategies that can help the sensor network run for extended time lapse must be investigated. Finally, biofouling, sensor drift, and underwater connectivity are all problems that should be considered in the implementation of a wireless sensor network for water quality monitoring. The overall efficiency and usefulness of wireless sensor network-based water quality monitoring systems can be enhanced by solving these problems.

References

1. Moqing D., Zhao, Y., Chen, S., 2012. "Automatic measurement and reporting system of water quality based on GSM". In: 2012 Second International Conference on Intelligent System Design and Engineering Application, pp. 1007–1010. IEEE
2. Li, S., Wang, H.; Xu, T. and Zhou, G. 2011. "Application study on Internet of Things in environment protection field", Lecture Notes in Electrical Engineering, vol. 133, pp. 99–106.
3. Lambrou P., Anastasiou C., Panayiotou G. and Polycarpou M., 2014. "A Low-Cost Sensor Network for Real-Time Monitoring and Contamination Detection in Drinking Water Distribution Systems". IEEE Sensors Journal, 2765–2772
4. FAO, "Chapter 2 - Water Quality Monitoring, Standard and Treatment,"[Online]. Available: http://www.fao.org/docrep/X5624E/x5624e05.htm.Accessed Jun. 18, 2017.
5. Andreea Mihaela dunce, 2018. "Water Pollution and Water Quality Assessment of Major Transboundary Rivers from Banat (Romania)", Volume 2018 |ArticleID 9073763 | https://doi.org/10.1155/2018/907376[3, 2018
6. Farrell-Poe K., 2018. "Water quality and monitoring". [Online]: https://cals.arizona.edu/watershedsteward/resources/docs/gui de/(10)Water%20Quality.pdf
7. Mompoloki Pule , Abid Yahya, Joseph Chuma, 2017. "Wireless sensor networks: A survey on monitoring water quality" , Journal of Applied Research and Technology, https://doi.org/10.1016/j.jart.2017.07.004 1665-6423
8. WiFi Direct 2015. NFC vs Bluetooth vs Wifi Direct: Comparison, Advantages and Disadvantages. (2015). Retrieved August, 2016 from http://www.itechwhiz.com/2014/01/nfc-bluetooth-Wifidirect-comparison-differences.htm
9. Rasin Z. and Abdullah M., 2009. "Water quality monitoring system using zigbee based wireless sensor network". International Journal of Engineering & Technology 9, 10 (2009), 24–28.
10. Adamo F., Attivissimo F.,Carducci C., and Lanzolla L., 2015." A Smart Sensor Network for Sea Water Quality Monitoring". Sensors Journal, IEEE 15, 5 (2015), 2514–2522.
11. Dhawan S., 2007. "Analogy of promising wireless technologies on different frequencies: Bluetooth, WiFi, and WiMAX". In Proc. Of the 2nd International Conference on Wireless Broadband and Ultra Wideband Communications. IEEE, Sydney, NSW, 14. DOI: https://doi.org/10.1109/AUSWIRELESS.2007.27
12. Lee C. and Wong K., 2010. "Planar monopole with a coupling feed and an inductive shorting strip for LTE/GSM/UMTS operation in the mobile phone". IEEE Transactions on Antennas and Propagation 58, 7 (2010), 2479–2483.
13. Alkandari A., Alnasheet M., Alabduljader Y., and Moein S.M., 2011. "Wireless sensor network (WSN) for water monitoring system: case study of Kuwait Beaches". International Journal of Digital Information and Wireless Communications (IJDIWC) 1, 4 (2011), 709–717.
14. Rao A.,Marshall S., Gubbi J., Palaniswami M., Sinnott R., and Pettigrovet V., 2013. "Design of low-cost autonomous water quality monitoring system". In Proc. of Advances in Computing, Communications and Informatics (ICACCI). IEEE, Mysore, India, 14–19. DOI: https://doi.org/10.1109/ICACCI.2013.6637139
15. Noreen, U., Bounceur, A., & Clavier, L. , 2017. "A study of LoRa low power and wide area network technology". In: 2017 International Conference on Advanced Technologies for Signal and Image Processing (ATSIP), (pp. 1–6), IEEE.
16. Minhai Zhang, Shuangxiang She, 2017. "Wastewater Monitoring System in Industrial Workshop Based on Wireless Sensor Network", https://doi.org/10.3991/ijoe.v13i03.6860
17. Abid Yahya, Joseph Chuma, 2017. " Wireless sensor networks: A survey on monitoring water quality", Journal of Applied Research and Technology,
18. Gungor, V. C., Hancke, G. P., & Member, S., 2009. Industrial wireless sensor networks: Challenges, design principles, and technical approaches. IEEE Transactions on Industrial Electronics, 56(10), 4258–4265.
19. Bhende, M., Wagh, S. J., & Utpat, A. , 2014. A quick survey on wireless sensor networks. In Proceedings – 2014 4th international conference on com- munication systems and network technologies, CSNT 2014 (pp. 160–167). https://doi.org/10.1109/CSNT.2014.40

20. Nabar S., Walling J., and Poovendran R., 2010. "Minimizing energy consumption in body sensor networks via convex optimization", in Proc. 2010 IEEE International Conf. Body Sensor Networks, pp. 62–67.
21. Lu X., Wang P., Niyato D., Kim D.I. and Han Z., 2015. "Wireless networks with RF energy harvesting: a contemporary survey", IEEE Communications Surveys & Tutorials, vol. 17, no. 2, pp. 757-789
22. M. Deqing, Z. Ying, C. Shangsong: Automatic measurement and reporting system of water quality based on GSM. In: 2012 Second International Conference on Intelligent System Design and Engineering Application, pp. 1007–1010. IEEE, 2012.
23. V. Chapmana, C. Bradleyb ,M. Gettelc , G. Hatvanid , T. Heine , J. Kovácsf , I. Liskag , M. Oliverh , P. Tanosi , B. Trásyf , G. Várbírój ; Developments in water quality monitoring and management in large river catchments using the Danube River as an example. Environmental Science & Policy, 64 (5), 141–154. doi:https://doi.org/10.1016/j.envsci.2016.06.015, 2016.
24. M. Nasirudin, U. Za'bah, O. Sidek, Fresh water real-time monitoring system based on wireless sensor network and GSM. 2011 IEEE Conference on Open Systems (ICOS2011), Sept. 25–28, 2011.
25. MS. BenSaleh, M. Qasim and AM. Obeid, A. Garcia-Ortiz, A Review on Wireless Sensor Network for Water Pipeline Monitoring Applications, Proceedings of the 2013 International Conference on Collaboration Technologies and Systems, DOI 10.1109/CTS.2013.6567217, 2013.
26. S. Menon, M. Ramesh, P. Divya, "A low cost wireless sensor network for water quality monitoring in natural water bodies", In IEEE Global Humanitarian Technology Conference, USA, 25 Dec. 2017.
27. M. Manju, V. Karthik, S. Hariharan, B. Sreekar, "Real time monitoring of the environmental parameters of an aquaponic system based on Internet of Things", Proc. of IEEE International conference on Science, Technology, Engineering and Management, India, 18 Jan. 2018
28. Yue, R., & Ying, T., 2012. A novel water quality monitoring system based on solar power supply & wireless sensor network. Procedia Environmental Sciences, 12(Icese 2011), 265 272. https://doi.org/10.1016/j.proenv.2012.01.276
29. Mo, Deqing., Zhao, Y., Chen, S., 2012. "Automatic measurement and reporting system of water quality based on GSM". In: 2012 Second International Conference on Intelligent System Design and Engineering Application, pp. 1007–1010. IEEE
30. Vijayakumar, N., & Ramya, R., 2015. The real time monitoring of water quality in IoT environment. In IEEE sponsored 2nd international conference on innovations in information, embedded and communication systems (ICIIECS) 2015 (pp. 1–4).
31. Aswale, P., Patil, S., Ahire, D., Shelke, S. and Sonawane, M., 2015. "Water environment monitoring system based on WSN", International Journal of Advanced Research in Electronics and Communication Engineering (IJARECE), Vol. 4, No. 4, p. 4.
32. Ranjbar, M.R., Abdalla, A.H. , 2017. "Low-cost, real-time, autonomous water quality testing and notification system". Int. J. Comput. Sci. Netw. Secur. 17(5), 277–282
33. Mabrouki, J., Azrour, M., Farhaoui, Y. and El Hajjaji, S., 2020. "Intelligent system for monitoring and detecting water quality", in Farhaoui, Y. (Ed.): Big Data and Networks Technologies, Vol. 81, pp.172–182, Springer International Publishing, Cham.
34. Sathish P, Gandla S. J., 2020. "Smart water quality monitoring system with cost-effective using IoT", https://doi.org/10.1016/j.heliyon.2020.e04096
35. Monira M., Surajit D. B. , Samia I., Ahmed W. R., Saddam H. K., 2020. "IoT based Smart Water Quality Monitoring System", IEEE 4th International Conference on Computer and Communication Systems , ISBN 9781728161365.
36. Vaishnavi V. Daigavane and Gaikwad M.A., 2020. "Water Quality Monitoring System Based on IOT", Advances in Wireless and Mobile Communications. ISSN 0973-6972 Volume 10, Number 5 (2017), pp. 1107–1116.
37. Amruta, M. K., & Satish, M. T., 2013. "Solar powered water quality monitoring system using wireless sensor network". In International multi-conference on automation, computing.

38. Khetre, A. C., & Hate, S. G. (2013). Automatic monitoring & reporting of water quality by using WSN technology and different routing methods. International Journal of Advanced Research in Computer Engineering &Technology, 2(12), 3255–3260.
39. Chung, W.-Y., & Yoo, J.-H. (2015). Remote water quality monitoring in wide area. Sensors and Actuators B: Chemical, 217, 1–7. https://doi.org/10.1016/j.snb.2015.01.072
40. Samsudin, S.I., Salim, S.I.M., Osman, K., Sulaiman, S.F., Sabri, M.I.A. (2018): A smart monitoring of a water quality detector system. Indones. J. Electr. Eng. Comput. Sci. 10(3), 951–958 (2018)
41. Alexander T. Demetillo, Michelle V. Japitana & Evelyn B. Taboada (2019) A system for monitoring water quality in a large aquatic area using wireless sensor network technology Sustainable Environment Research volume 29, Article number: 12
42. Kofi Sarpong Adu-Manu,1,2 Ferdinand Apietu Katsriku,2 Jamal-Deen Abdulai,2 and Felicia Engmann3, (2020), Smart River Monitoring Using Wireless Sensor Networks
43. Paul D., Vivian F. López B. and Diego H. , 2020. "Intelligent WSN System for Water Quality Analysis Using Machine Learning Algorithms: A Case Study (Tahuando River from Ecuador)", Remote Sens. 2020, 12, 1988; doi:https://doi.org/10.3390/rs12121988www.mdpi.com/journal/remotesensing.
44. Faustine, A., Mvuma, A. N., Mongi, H. J., Gabriel, M. C., Tenge, A. J., &Kucel, S. B. (2014, December). Wireless sensor networks for water quality monitoring and control within lake victoria basin: Prototype development. 281–290. https://doi.org/10.4236/wsn.2014.612027
45. Lambrou, T. P., Anastasiou, C. C., Panayiotou, C. G., & Polycarpou, M. M. (2014). A low-cost sensor network for real-time monitoring and contamination detection in drinking water distribution systems. IEEE Sensors Journal, 14(8), 2765–2772. https://doi.org/10.1109/JSEN.2014.2316414
46. Pradeepkumar, M., Monisha, J., Pravenisha, R., Praiselin, V. and Devi, K.S. (2016) 'The real time monitoring of water quality in IoT environment', Int. J. Innov. Res. Sci. Eng. Technol., Vol. 5, No. 3, pp.4419–4427.Ranjbar and Abdalla (2017)
47. Simitha K M, Subodh Raj M S (2019) ,« IoT and WSN Based Water Quality Monitoring System » Proceedings of the Third International Conference on Electronics Communication and Aerospace Technology [ICECA 2019] IEEE Conference Record # 45616; IEEE Xplore ISBN: 978-1-7281-0167-5
48. Marco Carminati 1,* , Andrea Turolla 2,* , Lorenzo Mezzera 1, Michele Di Mauro 2, Marco Tizzoni 2, Gaia Pani 2, Francesco Zanetto 1 , Jacopo Foschi 2 and Manuela Antonelli 2 (2019), A Self-Powered WirelessWater Quality Sensing Network Enabling Smart Monitoring of Biological and Chemical Stability in Supply Systems
49. Laiqa Binte Imran1 · Rana Muhammad Amir Latif1 · Muhammad Farhan1 · Hamza Aldabbas2 (2020) Smart City Based Autonomous Water Quality Monitoring System Using WSN ; Wireless Personal Communications , https://doi.org/10.1007/s11277-020-07655-x
50. Ayadi A., Ghorbel O., BenSalah M.S. , Abid M., 2020. "A framework of monitoring water pipeline techniques based on sensors technologies", Journal of King Saud University – Computer and Information Sciences, https://doi.org/10.1016/j.jksuci.2019.12.003
51. Wang, Z., Wang, Q., & Hao, X. (2009). The design of the remote water quality monitoring system based on WSN. In 5th international conference on wireless communications, networking and mobile computing, 2009. WiCom '09(pp. 1–4). https://doi.org/10.1109/WICOM.2009.5303974
52. Rao, A.S., Marshall, S., Gubbi, J., et al.: Design of low-cost autonomous water quality monitoring system. In: 2013 International Conference on Advances in Computing, Communications and Informatics (ICACCI), pp. 14–19. IEEE (2013)
53. Hongpin, L., Guanglin, L., Weifeng, P., Jie, S., Qiuwei, B. (2015): Real-time remote monitoring system for aquaculture water quality. Biol. Eng. 8, 8 (2015)
54. Parra, L., Sendra, S., García, L. and Lloret, J. (2018) 'Design and deployment of low-cost sensors for monitoring the water quality and fish behavior in aquaculture tanks during the feeding process', Sensors, March, Vol. 18, No. 3, p.750.

Simulation of the Treatment Performance of a Purification Plant for a Dairy Effluent

Toufik Rachiq, Jamal Mabrouki, Souad El Hajjaji, and Sabir Rahal

1 Introduction

Water represents the most important raw material on our planet, for humans, animals, plants, and microorganisms [1]. Practically, all vital phenomena in the biosphere are linked to the availability of water. Water is not only a living space, an energy carrier or a means of transportation, but also an essential element for any kind of production [2].Therefore, they should be directed to wastewater treatment plants whose role is to concentrate the pollution contained in the wastewater in the form of a residue and to discharge a purified water that meets the accepted standards, and this through physicochemical and biological processes [3].

The decontamination of urban wastewater requires a succession of steps involving physical, physicochemical, and biological treatments [4]. Apart from the largest waste present in the wastewater, the treatment must allow, at the very least, to eliminate most of the carbonaceous pollution. Some processes even allow the elimination of nitrogen from phosphorus. A large majority of these pollutants are transferred from the liquid phase to a concentrated sludge phase [5].

T. Rachiq (✉)
Laboratory of Spectroscopy, Molecular Modeling, Materials, Nanomaterial, Water and Environment, CERNE2D, Mohammed V University in Rabat, Faculty of Science, Agdal, Rabat, Morocco

Central Dairy Plant, Sale, Morocco

J. Mabrouki · S. El Hajjaji
Laboratory of Spectroscopy, Molecular Modeling, Materials, Nanomaterial, Water and Environment, CERNE2D, Mohammed V University in Rabat, Faculty of Science, Agdal, Rabat, Morocco

S. Rahal
Central Dairy Plant, Sale, Morocco

M. Azrour et al. (eds.), *IoT and Smart Devices for Sustainable Environment*, EAI/Springer Innovations in Communication and Computing,
https://doi.org/10.1007/978-3-030-90083-0_2

The activated sludge process is the most used process in Morocco to treat wastewater. Although the purification performance and reliability of this process is approved, several types of malfunctions may occur [6]. The most frequent is the excessive development of filamentous bacteria, which can lead to a degradation of the sludge settling or a stable foaming [7]. The protection of the environment is currently one of the priority objectives of our country [8]. Watercourses are affected by punctual and chronic pollution. The rivers are affected by point and chronic pollution. These effects result mainly from the activity of industries, farms, and urban wastewater [9]. The consequences of these liquid discharges on ecosystems are sometimes very serious [10]. As a result, today, many countries have adopted more or less strict standards that set thresholds for pollution discharges that must not be exceeded.

The operation of a WWTP always finds it difficult to adopt an operating regime, given the daily fluctuations in pollutant load (COD, BOD, TSS etc. . . .) and hydraulic load (volumes of wastewater arriving at the WWTP), although the water discharged meets the effluent standard requirements. It is in this context that a numerical simulation of the operating regime of the WWTP by the GPS-X 7 software [11], taking into account the physical and operational parameters of the plant, was carried out to determine the optimal operating conditions and save operators a lot of time in adapting an operating regime regardless of variations in flows and loads arriving at the WWTP. Our main goal in this paper is to obtain the optimal operation of the WWTP from a GPS-X software and calibrates the WWTP data by the results obtained by GPS-X.

2 Materials and Methods

2.1 Parameter Analysis

The reactions govern the fate of the various variables that make up the effluent use stoichiometric and kinetic parameters. Depending on the process, we find stoichiometric parameters that illustrate the yield of heterotrophic and autotrophic bacteria, the fraction of biomass that is transformed into a particulate product, and the proportion of nitrogen contained in the biomass and in the products of its decomposition. Similarly, kinetic parameters illustrate the growth and death of bacteria, ammonification, hydrolysis, and correction factors in anoxic conditions for heterotrophic growth and hydrolysis. A total of five stoichiometric and kinetic parameters are available.

2.2 Software Uses

GPS-X, designed since 1988, runs on a PC under the Windows operating system. Although it is marketed only in English, it is used all over the world by consulting companies, industries, and municipalities. It is constantly evolving and adapting to research in the field of sanitation. The software integrates many models listed in a library. The models used are both internationally published models (ASM) and models developed internally by Hydromantis. In addition, the user can create his own model [12]. The advantage of the GPS-X (global purpose system) software is that it is very powerful and user-friendly, provided that you master the different sound possibilities and operation. However, it is in constant evolution according to the needs of the market. By elsewhere, its effectiveness is linked to the use of a large amount of data. Simulations can be carried out in both steady state and steady-state operation, dynamic. As an example, Makinia [13] performed steady-state simulations from a set of urban wastewater fractionation plants in northern Poland, and the acquired coefficients have been validated in dynamic regime. In each regime, it is possible to make calibrations, adjustments, sensitivity analyses, and process customization. Nitrification and denitrification efficiencies are obtained for different stations. The programming language of the software used is FORTRAN and ACSL (Advanced Continuous Simulation Language; java interface) [14]. The software allows us to create the current configuration of the station. The use of GPS-X is done by creating simplified diagrams of the station using predefined tools. Each tool represents a structure (inlet effluent, aeration basin, clarifier, etc.) that the designer must then assemble to define the hydraulic profile. The GPS-X is a tool that allows us to configure any type of operation, so it is an aid for simulation, analysis, and optimization [15].

2.3 Model Selection

The ASM1 model is combining a simple description of biological phenomena and a representation in conformity with reality. It expresses the degradation rates of the carbonaceous substrate and nitrogen depending on the state of pollution (named by the variables) and characteristics of the biomass providing the treatment (named by parameters). The ASM1 model uses different notions that we will try to describe [16]. An activated sludge system includes phenomena such as the oxidation of the carbon, nitrification, and denitrification. The simulation of its behavior involves numerous reactions between a large number of components. To be mathematical while providing realistic predictions, the responses must represent the most relevant information available, fundamental processes of the system. In addition, the model must take into account the kinetics (reaction speed) and stoichiometry (ratio that one component has to the others in a chemical reaction) of each process [17]. The main conceptual tasks in the development of the mathematical model are the identification of the reactions and the choice of their appropriate kinetic and stoichiometric expressions. Combining a simple description of the biological phenomena and a representation conforming to reality, they express the rates of degradation of carbonaceous and nitrogenous substrates as a function of the state of pollution and the characteristics of the biomass providing the treatment. The ASM1 model uses different notions that we will try to describe and characteristics of the biomass providing the treatment [18].

3 Results and Discussions

In order to comply with the provisions of the water law, to better manage investments and in order to help operators better manage and control the water supply and waste water. To operate wastewater treatment plants (WWTPs), the computer modeling tool has become indispensable. The latter is necessary for the knowledge of the mechanisms of the WWTP. The biological and hydraulic simulations of the different processes of the WWTP in order to obtain an optimal solution from the point of view of purification performance and to offer indispensable elements to enable the environment ministry officials to better scientifically and technically base their management of the treatment plants [19] (Fig. 1).

3.1 *Input Effluent Characteristics*

Before starting the simulation phase of the treatment process, we studied the data from the four-month assessment carried out in 2017 by the Ministry of the Environment during this treatment plant.

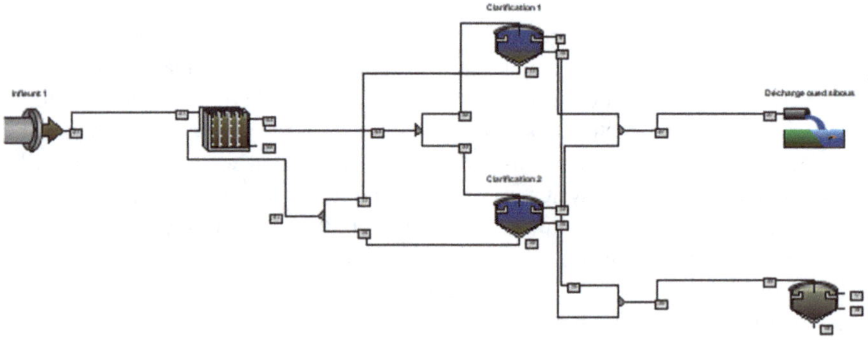

Fig. 1 Diagram of a WWTP in GPS-X

Table 1 The characteristics of the input effluent

Parameter	NS	Minimum	Average	Maximum	ST
pH	16	2.5	8.54	11.9	2.807
T(C°)	16	26	28.2	31	1.939
Cond(μs/cm)	16	4110	1630	8300	2432.37
Tur (NTU)	16	626	761.066	1030	166.01
TSS(mg O2/l)	16	288	510.571	718	146.22
COD(mgO2/l)	16	2317	3753.785	5492	1048.512
BOD5(mgO2/l)	16	1218	1994.064	2591	499.346
NTK(mgO2/l)	16	30	52.25	72	14.691
BOD5/COD	16	0.525	0.531	0.472	0.566
SS/BOD5	16	0.236	0.256	0.277	0.139

Table 1 presents the results of the main physicochemical and bacteriological characteristics of the overall raw effluent discharged by the studied dairy. The values recorded for the parameters TSS, COD, BOD5, N-NTK, and P-PT, as well as the pH and temperature values, largely exceed the values set by the Moroccan draft standards of limit values for liquid discharges. Indeed, the pH of the effluent analyzed varies greatly, with sometimes very acidic (2.5) or very basic (11.9) values in relation to the use of nitric acid or soda for washing. The same applies to the temperature, which sometimes reaches 31°C as a result of direct discharge of water from the refrigeration or barometric condensers. The annual organic load is on average 3753.7 mg O2/L for COD and 1994.1 mg O2/L for BOD5.

3.2 *Simulation of WWTP Operations*

The simulation consists in performing virtual experiments on the operation mode of the WWTP, using the GPS-X software and one or more equation models to obtain the optimal operation of the WWTP. The objectives of these simulations are the enrichment of knowledge and a better understanding of the activated sludge system of the WWTP. This is done by testing different scenarios in order to anticipate and predict the variations that can be generated and finally to optimize the operation of this process [20].

The calibration of a model consists in adjusting its parameters in order to simulate the most accurate simulation of the process of the real system. The model encompasses a large number of parameters, and it would be tedious to modify each parameter individually in order to adjust the values simulated on the measured values [21, 22]. In addition, some parameters are linked, their influence on reaction kinetics is directly proportional, and it is necessary to consider the combination of parameters that they constitute. It is therefore necessary to define a precise method for calibrating the model, in particular to reduce the number of parameters to be

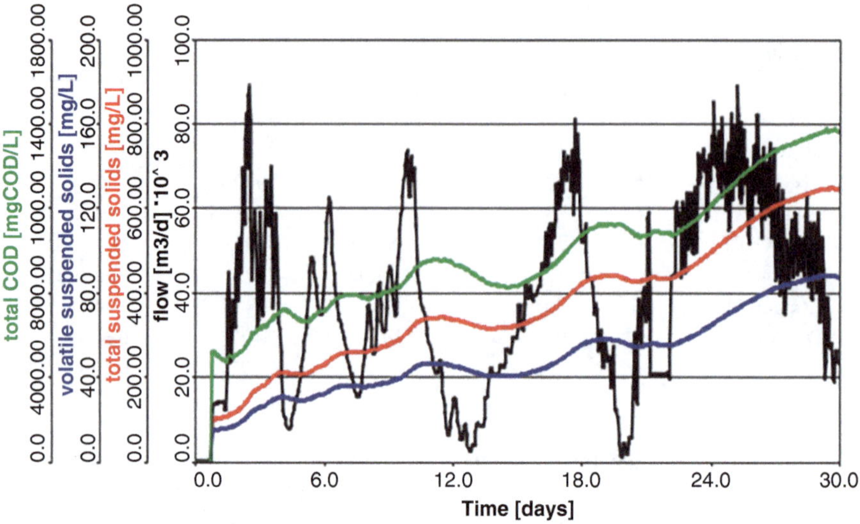

Fig. 2 Curve displayed by the software for one month 1

modified and to determine the sequences of measured values that will be used to calibrate the parameters.

The calibration of a model can be done visually or mathematically, by a simulation algorithm. Given the complexity of the ASM1 model and the lack of detailed data required by the automatic calibration, it is not possible to carry out directly a mathematical simulation for practical identification problems of the parameters [23]. We have therefore opted for visual and manual adjustment of the parameters on the experimental measurements. The results are displayed as an Excel file, containing curves representative of the different parameters (flow m^3 /d, TSS, BOD5 and COD in mg/l) (Fig. 2).

In order to address the question of how to achieve optimal WWTP performance using GPS-X software, we must first ascertain the GPS-X software; we will first have to ensure the purification efficiency of the WWTP by activated sludge. In order to achieve our goal, we spent 01 months of practical training at the wastewater treatment plant with the agreement of the water board and our hydraulic department. During our internship, we followed almost daily the quality of the water through the output work, which is the clarifier by following the parameters available at the WWTP. Then, we proceeded to the calibration between the observed values and the values calculated from GPS-X.

The COD concentrations at the outlet obtained by the analyses vary from 652 to 857 mg/L, while the COD concentrations simulated by GPS-X vary from 457 to 562 mg/L; these results are above the results that still exceed the discharge standard, which is set at 100 mg/L. The BOD_5 concentrations at the outlet obtained by the analyses vary from 156 to 269 mg/L; while the COD concentrations simulated by GPS-X vary from 156 to 214 mg/L, these results meet the standard. The amount of

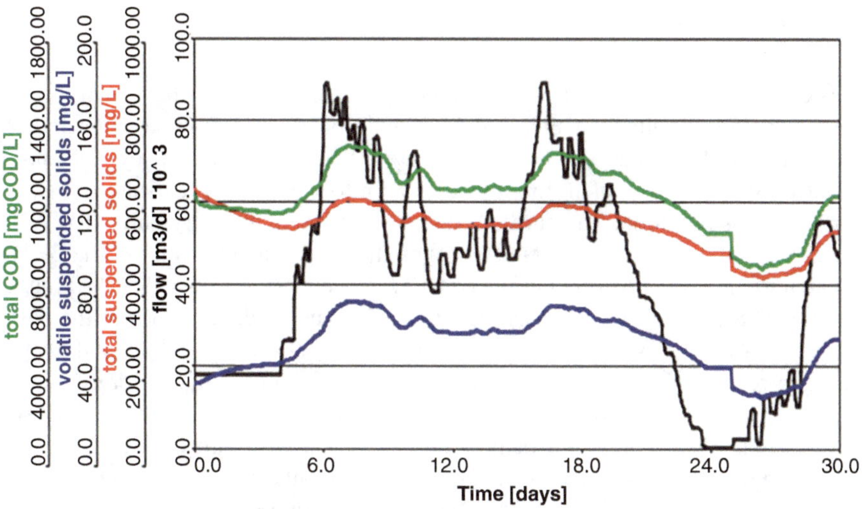

Fig. 3 Curve displayed by the software for month 2

suspended matter at the inlet varies between 172 mg/l and 400 mg/l, with an average of 316.07 mg/l, and at the outlet we obtained a minimum value of 110 mg/l and a maximum value of 210 mg/l, with an average of 160 mg/l; this value is higher than the standard of the discharge applied in Morocco (50 mg/l). The values of TSS at the entrance of the station vary between 288 mg/l and 718 mg/l, with an average of 510.6 mg/l. This variation is due to instability of the quality of water at the entrance. At the exit of the plant, we record a value of TSS 16–28 mg/l with an average of 22 mg/L; this value is in accordance with the discharge standard. Therefore, the TSS removal efficiency exceeds 60% (Fig. 3).

The COD concentration values of the raw water of the WWTP vary between 2317 and 5492 mg/1 with an average of 3753.7. On the other hand, the values of the concentrations of the treated water vary between a maximum of 510 mg/1 and a minimum of 657 mg/1 and a COD removal efficiency of about 65%. These values do not comply with the Moroccan discharge standard (<500 mg/l), so a large concentration has been eliminated by this process because there is an important degradation of the pollutant load. The curve represents the variation of the BOD5 of the raw water and the treated water; let us note that the raw water at the entrance of the plant presents a BOD5, which varies between 1218 and 2591 mg/L. On the other hand, for the treated water, we notice that the BOD5 at the outlet 190.45 and 160.66 mg/l of these values are generally all within the standards of rejection (lower than 50 mg of O2/l). After several simulation scenarios, the optimal operating model close to the analysis results is obtained for the results is obtained for the values of the recirculation and extraction flows, respectively, 2543 m^3/day and 562 m^3/day.

4 Conclusion

Wastewater treatment by activated sludge processes relies on the activity of an aerobic bacterial culture, maintained in suspension in a specific structure fed by the effluent to be treated and called an aeration tank. Aeration is a key element of a wastewater treatment plant to provide oxygen to the microorganisms to degrade carbonaceous pollution. The purified wastewater leaving the dairy effluent WWTP meets the requirements of the standard set by the Moroccan regulations. The BOD5 concentration is less than 50 mg/L, the COD concentration is less than 100 mg/L, and the TSS concentration is less than 30 mg/L. The modeling work of the dairy effluent WWTP uses the GPS-X software by the ASM1 model (including the nitrogen removal processes). The ASM1 model is the most suitable for our study. To model in proportion to an activated sludge system performing carbon oxidation, nitrification, nitrification, and denitrification, 13 variables are considered as the minimum required. Associated with the four fundamental processes (growth, death, ammonification, and hydrolysis), kinetic equations are proposed to bring into play all the biological relationships of the system through a calculation matrix. Moreover, the model presents a method to evaluate all the method, all the variables, and parameters necessary to solve the proposed equations. The study of the purification performances of the wastewater plant of Sale-Rabat by the analysis of the pollution criteria (BOD5, TSS, and COD) is made according to the standards in force. Our study showed that the values found respect the standards of rejection at the exit of the WWTP. The removal of COD, BOD5, and TSS with respective close yields for the years 2017 and 2018 (65%, 72%, and 65% for 2017 and 62%, 82.5%, 45% for 2018).

References

1. Allen, M. J., Edberg, S. C., & Reasoner, D. J. (2004). Heterotrophic plate count bacteria—what is their significance in drinking water? International journal of food microbiology, 92(3), 265–274.
2. Züttel, A., Remhof, A., Borgschulte, A., & Friedrichs, O. (2010). Hydrogen: the future energy carrier. Philosophical Transactions of the Royal Society A: Mathematical, Physical and Engineering Sciences, 368(1923), 3329–3342.
3. Singh, K. P., Mohan, D., Sinha, S., & Dalwani, R. (2004). Impact assessment of treated/untreated wastewater toxicants discharged by sewage treatment plants on health, agricultural, and environmental quality in the wastewater disposal area. Chemosphere, 55(2), 227–255.
4. Samal, K., Dash, R. R., & Bhunia, P. (2017). Treatment of wastewater by vermifiltration integrated with macrophyte filter: a review. Journal of environmental chemical engineering, 5(3), 2274–2289.
5. Lefebvre, O., & Moletta, R. (2006). Treatment of organic pollution in industrial saline wastewater: a literature review. Water research, 40(20), 3671–3682.
6. Shukla, A. A., Hubbard, B., Tressel, T., Guhan, S., & Low, D. (2007). Downstream processing of monoclonal antibodies—application of platform approaches. Journal of Chromatography B, 848(1), 28–39.

7. Nielsen, P. H., Kragelund, C., Seviour, R. J., & Nielsen, J. L. (2009). Identity and ecophysiology of filamentous bacteria in activated sludge. FEMS microbiology reviews, 33(6), 969–998.
8. Inglehart, R. (1995). Public support for environmental protection: Objective problems and subjective values in 43 societies. PS: Political science and politics, 28(1), 57–72.
9. Keraita, B., Drechsel, P., & Amoah, P. (2003). Influence of urban wastewater on stream water quality and agriculture in and around Kumasi, Ghana. Environment and Urbanization, 15(2), 171–178.
10. Bormann, F. H., & Likens, G. E. (2012). Pattern and process in a forested ecosystem: disturbance, development and the steady state based on the Hubbard Brook ecosystem study. Springer Science & Business Media.
11. Mabrouki, J., Benbouzid, M., Dhiba, D., & El Hajjaji, S. (2020). Simulation of wastewater treatment processes with Bioreactor Membrane Reactor (MBR) treatment versus conventional the adsorbent layer-based filtration system (LAFS). International Journal of Environmental Analytical Chemistry, 1–11.
12. Hauduc, H., Rieger, L., Oehmen, A., Van Loosdrecht, M. C. M., Comeau, Y., Héduit, A., ... & Gillot, S. (2013). Critical review of activated sludge modeling: state of process knowledge, modeling concepts, and limitations. Biotechnology and Bioengineering, 110(1), 24–46.
13. Makinia, J., Rosenwinkel, K. H., & Spering, V. (2005). Long-term simulation of the activated sludge process at the Hanover-Gümmerwald pilot WWTP. Water research, 39(8), 1489–1502.
14. Sánchez, J., Morilla, F., Dormido, S., Aranda, J., & Ruipérez, P. (2002). Virtual and remote control labs using Java: a qualitative approach. IEEE Control systems magazine, 22(2), 8–20.
15. Jeppsson, U., & Pons, M. N. (2004). The COST benchmark simulation model—current state and future perspective.
16. Iacovides, H., Launder, B. E., & Li, H. Y. (1996). The computation of flow development through stationary and rotating U-ducts of strong curvature. International Journal of Heat and Fluid Flow, 17(1), 22–33.
17. Smolders, G. J. F., Van der Meij, J., Van Loosdrecht, M. C. M., & Heijnen, J. J. (1995). A structured metabolic model for anaerobic and aerobic stoichiometry and kinetics of the biological phosphorus removal process. Biotechnology and Bioengineering, 47(3), 277–287.
18. Matsumura, Y., Sasaki, M., Okuda, K., Takami, S., Ohara, S., Umetsu, M., & Adschiri, T. (2006).Supercritical water treatment of biomass for energy and material recovery. Combustion Science and Technology, 178 (1–3), 509–536.
19. Langergraber, G., Rieger, L., Winkler, S., Alex, J., Wiese, J., Owerdieck, C., ... & Maurer, M. (2004). A guideline for simulation studies of wastewater treatment plants. Water Science and Technology, 50(7), 131–138.
20. Gernaey, K. V., van Loosdrecht, M. C., Henze, M., Lind, M., & Jørgensen, S. B. (2004). Activated sludge wastewater treatment plant modelling and simulation: state of the art. Environmental modelling & software, 19(9), 763–783.
21. Naessens, W., Maere, T., & Nopens, I. (2012). Critical review of membrane bioreactor models–Part 1: Biokinetic and filtration models. Bioresource Technology, 122, 95–106.
22. Mabrouki, J., Moufti, A., Bencheikh, I., Azoulay, K., El Hamdouni, Y., & El Hajjaji, S. (2019, July). Optimization of the Coagulant Flocculation Process for Treatment of Leachate of the Controlled Discharge of the City Mohammedia (Morocco). In International Conference on Advanced Intelligent Systems for Sustainable Development, pp. 200–212.
23. Sin, G., De Pauw, D. J., Weijers, S., & Vanrolleghem, P. A. (2008). An efficient approach to automate the manual trial and error calibration of activated sludge models. Biotechnology and bioengineering, 100(3), 516–528

Study of Piroxicam Removal from Wastewater by Artichoke Waste Using NemrodW® Software: Statistical Analysis

Nora Samghouli, Imane Bencheikh, Karima Azoulay, Fatima-Zahra Abahdou, Jamal Mabrouki, and Souad El Hajjaji

1 Introduction

Pharmaceuticals are active and stable compounds produced to boost human beings' health [1]. In recent decades, concerns over the presence of pharmaceuticals in the environment have increased, because of their presence in treated wastewater also because of their toxicity and antibacterial resistance [1, 2]. Water pollution is caused by the excretion of cosmetics, pharmaceuticals, and their metabolites, although it may also lead to improper disposal of unused and expired products [3, 4]. The contaminations of water systems by drugs are attributed mainly to the treatment inefficiency inside the wastewater treatment plants (WWTP). Since these purification facilities were not designed to degrade these categories of pollutants [5]. Although there are several possible tertiary treatment methods for pharmaceuticals active compounds (PhACs) removal, such as reverse osmosis, photo-degradation, chemical oxidation, coagulation, or ultraviolet radiation. However, these techniques are not very used in this kind of treatment, since they generate high quantities of wastes also have high operational costs [5–7]. Adsorption is considered among the simplest and most efficient techniques that can be applied as a tertiary treatment in WWTP for drug removal. It is characterized by high efficiency; also it is an eco-friendly method as it does not generate toxic byproducts. It is generally carried out with activated carbon characterized by its high adsorption capacity and its abundant porous structure. However, it has the disadvantage of high cost and high energy consumption. In this context, studies have shown that other materials can

N. Samghouli (✉) · I. Bencheikh · K. Azoulay · F.-Z. Abahdou · J. Mabrouki · S. El Hajjaji (✉)
Laboratory of Spectroscopy, Molecular, Modelling, Materials, Nanomaterials, Water and Environment, (LS3MNWE), Department of Chemistry, Faculty of Sciences, Mohammed V University in Rabat, Rabat, Morocco
e-mail: nora_samghouli@um5.ac.ma; souad.elhajjaji@um5.ac.ma

M. Azrour et al. (eds.), *IoT and Smart Devices for Sustainable Environment*, EAI/Springer Innovations in Communication and Computing,
https://doi.org/10.1007/978-3-030-90083-0_3

be used as an economical alternative to activated carbon [7, 8]. Among these materials, there are the lignocellulosic food and agricultural residues, that are used as absorbents for pollutants in aqueous effluents [9–11] due to their physical and chemical properties such as the presence of functional groups on their surface that facilitate the adsorption process, and their containment of a high specific surface area [12, 13]. In this study, we have chosen artichoke wastes as an alternative adsorbent to eliminate the anti-inflammatory piroxicam drug in an aqueous medium. Artichoke waste has been used in the adsorption of different organic pollutants such as dyes [14–17] and drugs [18].

The use of IT (information technology) software is part of the desire to reduce prices and time. For the optimization of all adsorption factors, functional and cost restrictions, we apply predictive analytics studies [19]. The aim is to obtain the key parameters that affect the adsorption efficiency and the estimated optimum conditions for the various factors. The determination of the significant parameters for the adsorption mechanism using the classical approach requires changing one factor at a time while setting all other factors at one stage and studying the variable's effect on the response. This operation has to be repeated as much as we have factors that result in a huge number of experiments. Therefore, for a multivariable operation, this is an incredibly time-consuming, costly, and complicated operation [20]. To solve this problem, statistical experimental design techniques use full factorial design for the study of parameters affecting the piroxicam removal by artichoke waste.

2 Materials and Methods

2.1 Preparation of Artichoke Waste

The artichoke waste has been chemically treated. The washed, dried, and ground artichoke waste was placed in a flask containing NaOH solution (2%). The flask was stirred and heated at 50 °C for 2 h in a water bath. The sample obtained was filtered and washed with distilled water and then dried. Then, treatment with sodium hypochlorite (NaClO) (3.7% w/v) was carried out at 70 °C in a ratio (0.5: 3 v/v) NaClO/H2O. The mixture was treated for 1h; subsequently the recovered sample was filtered and washed several times (until pH = 7), then air dried [16].

2.2 Preparation of Piroxicam Solutions

Piroxicam (Fig. 1) is an anti-inflammatory drug that ends up in wastewater via pharmaceutical industry activities and human consumption. This medicine is used

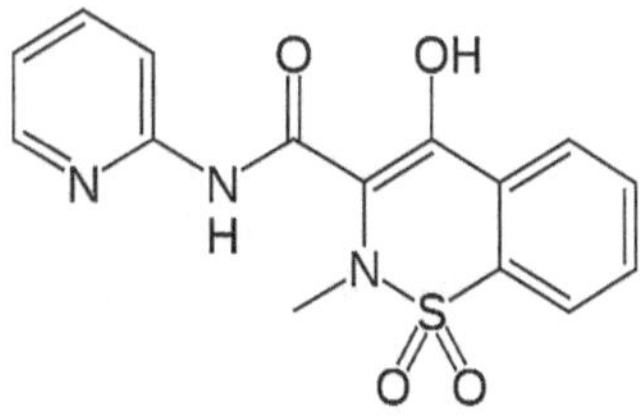

Fig. 1 Chemical structure of piroxicam

Table 1 Physical and chemical properties of piroxicam

Physical properties	Chemical properties			
Fusion temperature (°C)	Water solubility ($mg.L^{-1}$) at 22 °C	Chemical formula	Molar mass (g/mol)	pKa
198–200	23	$C_{15}H_{13}N_3O_4S$	331.346 ± 0.02	6.3

taken as a model pollutant drug in this study without any prior purification. Table 1 summarizes the physicochemical properties of piroxicam.

A stock solution has been prepared by dissolving 100 mg of the drug in 100 ml of distilled water. From the stock solution, four diluted solutions were prepared. The adsorption experiments are carried out by putting the dried and cleaned artichoke waste used as an adsorbent in a beaker containing 50 ml of piroxicam solution. The samples were then filtered, and the absorbance measurements were performed using a UV/visible spectrophotometer at 334 nm.

The adsorption efficiency is determined by the calculation of the piroxicam adsorption rate, noted R(%) using the following formula (1) [11]:

$$R\ (\%) = \frac{Ci - Ce}{Ci} \times 100 \tag{1}$$

Where C_i is the initial concentration (mg/L) and C_e is the equilibrium concentration (mg/L).

2.3 *Statistical Analysis: Determination of Piroxicam Key Adsorption Parameters*

The experimental designs are part of the general approach aiming to improve the efficiency of drug removal. The goal of the statistical analysis is to determine the significant factors that influence the piroxicam removal using the artichoke waste. The analysis is based on the investigation and the study of the different mathematical constants such correlation coefficient, p-Value, estimations constant, etc.

The modified linear determination coefficient ($R^2{}_{Adj}$) i n he Eq. 2 explains the variability measure of the linear model [21].

$$R^2_{Adj} = 1 - \frac{\text{SSResidual}/\text{DFResidual}}{(\text{SSModel} + \text{SSResidual}) / (\text{DFModel} + \text{DFResidual})} \quad (2)$$

Where SS is the sum of squares and DF is the freedom degrees.

Analysis of variance (ANOVA) was used to assess if the regression model is capable of reflecting experimental data; this was also checked by the appropriate accuracy ratio (Eq. (3)) [22, 23].

$$\text{Adequate precision} = \frac{\text{Max}\left(\hat{Y}\right) - \text{Min}\left(\hat{Y}\right)}{\sqrt{\tilde{V}\left(\hat{Y}\right)}} = \frac{\text{Max}\left(\hat{Y}\right) - \text{Min}\left(\hat{Y}\right)}{\sqrt{\frac{p\sigma^2}{n}}} \quad (3)$$

- $\sqrt{\tilde{V}\left(\hat{Y}\right)}$ is the medium variance of the predicted values, p is the model parameters number (β_0 is included), σ^2 is the residual mean square, and n is the number of experiments.

This gives a detailed interpretation of the piroxicam adsorption behavior with minimal experiments.

In the present case and as mentioned before, the statistical analysis determines the possible "significant" or "actives" variables' effects and appeals to the implementation of "statistical test" well-defined procedures. The analysis methods of Lenth and Henry are presented in this study.

- Lenth Method

The Lenth method [24] consists of estimating a pseudo-standard error, to implement a statistical test whose result is expressed in a graphical form similar to a control chart.

Active effects refer to parameters whose approximate values are below the bounds (Fig. 2). Conversely, inactive results refer to parameters whose approximate values are below the limits (Fig. 2).

- Henry Line Plot

Henry's method is presented depending on the distribution of the results (bj or |bj|): the normal probability plot (bj) [25] and half-normal plot (|bj|) [26, 27].

Fig. 2 Active and nonactive effects parameters

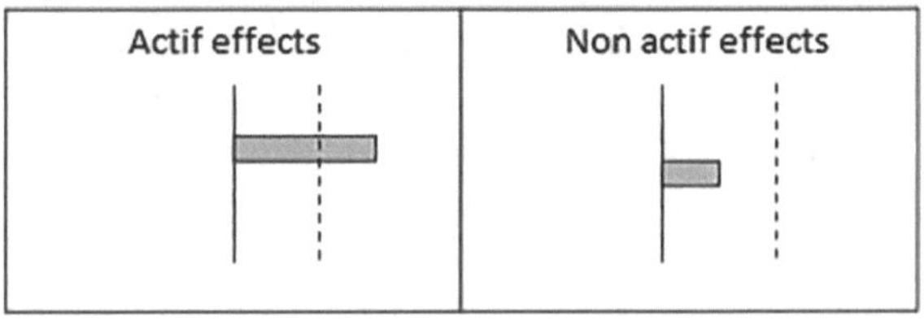

Table 2 Experimental area for the 2^4 facorial design

Factor	Symbol	Unit	Coded level	
			−1	+1
			Level value	
pH	(x_1)	–	3	8
Mass	(x_2)	g	0.5	2
Time	(x_3)	min	15	60
$[\text{Piroxicam}]_i$	(x_4)	mg/L	10	40

Table 3 Experimental matrix with the coded variables for the piroxicam adsorption via artichoke waste

Exp.	X_1	X_2	X_3	X_4
1	−1.0000	−1.0000	−1.0000	−1.0000
2	1.0000	−1.0000	−1.0000	−1.0000
3	−1.0000	1.0000	−1.0000	−1.0000
4	1.0000	1.0000	−1.0000	−1.0000
5	−1.0000	−1.0000	1.0000	−1.0000
6	1.0000	−1.0000	1.0000	−1.0000
7	−1.0000	1.0000	1.0000	−1.0000
8	1.0000	1.0000	1.0000	1.0000
9	−1.0000	−1.0000	−1.0000	1.0000
10	1.0000	−1.0000	−1.0000	1.0000
11	−1.0000	1.0000	−1.0000	1.0000
12	1.0000	1.0000	−1.0000	1.0000
13	−1.0000	−1.0000	1.0000	1.0000
14	1.0000	−1.0000	1.0000	1.0000
15	−1.0000	1.0000	1.0000	1.0000
16	1.0000	1.0000	1.0000	1.0000

The parameters' effects that deviate from the line on the normal plot (below or above the line, depending on whether the effects are positive or negative) are considered as "probably active." The half-normal plot makes it possible to better distinguish the active coefficients, but i t does not provide their significance since it considers here the coefficient absolute value.

The first step of the statistical study consists of choosing controllable parameters and the experimental field by determining the variation limits of each factor. To evaluate the variables (factors) that affect significantly the piroxicam removal performance, a full factorial design 2^4 is chosen (with 2 and 4 are, respectively, the number of the studied levels and factors). Adsorbent mass, pH, time, and initial concentration were the selected factors for the study. Table 2 indicates the ranges and levels of the variables (low (−1) and high (+1)) analyzed in this study. All experiments were conducted in random order, and the calculations were acquired by NemrodW® software [28].

The used experimental design is a full factorial design with four factors X_1, X_2, X_3, and X_4. This results in a matrix consisting of 16 experiments (2^4) [22]. Table 3 presents the experimental matrix with the coded variables for the piroxicam adsorption via artichoke waste.

As we used a full factorial design 2^4 with interactions, the polynomial equation is as follows [29]:

$$R(\%) = b_0 + b_1X_1 + b_2X_2 + b_3X_3 + b_4X_4 + b_{12}(X_1X_2) + b_{13}(X_1X_3) + b_{23}(X_2X_3) + b_{14}(X_1X_4) + b_{24}(X_2X_4) + b_{34}(X_3X_4) + b_{123}(X_1X_2X_3) + b_{124}(X_1X_2X_4) + b_{134}(X_1X_3X_4) + b_{234}(X_2X_3X_4) + b_{1234}(X_1X_2X_3X_4)$$

3 Results and Discussion

3.1 Statistical Analysis

The experimental matrix obtained by the full factorial design 2^4 and the obtained responses (effectiveness of piroxicam removal) are shown in Table 4. We observe from Table 4 that the piroxicam adsorption removal ranges from 5 to 42%.

Table 4 The experimental matrix of full factorial design 2^4 with the retention rate (%)

Exp.	pH	mass (g)	Time (min)	$[\text{Piroxicam}]_i$ (mg/L)	Retention rate (%)
1	3	0.5	15	10	18
2	8	0.5	15	10	17
3	3	2	15	10	26
4	8	2	15	10	26
5	3	0.5	60	10	18
6	8	0.5	60	10	37
7	3	2	60	10	42
8	8	2	60	10	38
9	3	0.5	15	40	21
10	8	0.5	15	40	10
11	3	2	15	40	11
12	8	2	15	40	5
13	3	0.5	60	40	29
14	8	0.5	60	40	21
15	3	2	60	40	30
16	8	2	60	40	22

3.2 *Determination of the Key Parameters for Piroxicam Adsorption via Artichoke Waste*

After the evaluation of each factor effect, we will select those that have a significant effect on the response, based on the effects plot and Pareto plot. The values of the factor effects are shown in Table 5, and the effect of the studied parameters is given by the coefficients represented in Fig. 3.

According to the effects graph, we notice that the initial piroxicam concentration, the pH-concentration interaction, and the mass-concentration interaction have an important effect on the retention rate. On one hand, the increase of these terms causes the decrease of the piroxicam adsorption percentage. On the other hand, the augmentation of the contact time contributes to the improvement of piroxicam removal via artichoke waste. The other factors' interactions have very negligible influences on the response. The coefficient b_3 related to the time factor is likely to prove the compatibility with the Lenth method.

3.3 *Pareto Analysis*

The Pareto analysis is another strategy used to show the importance of the parameters studied. This diagram shows the percentage of each term for the development of the Y response [20].

To facilitate the determination of influencing factors, we have applied the Pareto method whose principle is that 20% of the parameter acts on 80% of the adsorption yield.

This graph represents Pareto graphical analysis that illustrates the key factors for the piroxicam adsorption (Fig. 4).

Pareto diagram analysis (Fig. 4) shows that time, the initial piroxicam concentration, and the pH-concentration interaction, with the mass-concentration interaction, have an important effect on 80% of piroxicam adsorption percentage, which confirms the results obtained by the effects graph.

From Pareto graphic analysis, the parameters above 20% have been removed. The equation of the adopted mathematical model will be as follows:

$$\begin{aligned} R\,(\%) = {} & 23.2 + 6.41X_3 - 4.53X_4 - 2.97X_1X_4 - 3.34X_2X_4 + 1.7X_1X_2X_4 + 1.8X_2 \\ & - 1.8X_1X_2X_3 + 1.6X_2X_3 - 1.2X_1 + 1.2X_1X_2X_3X_4 - 1.1X_1X_2 + 1.1X_1X_3 \\ & - 0.9X_1X_3X_4 + 0.4X_3X_4 + 0.6X_2X_3X_4 \end{aligned}$$

The analysis of variance (ANOVA) is presented in Table 6, and the results show the correlation between the experimental and calculated results since the p-value is less than 5% [30]. The coefficient of regression of piroxicam adsorption by artichoke waste is $R^2 = 0.9363$ and is near to 1, allowing us to conclude that 93.63% of the studied phenomenon is explained by the theoretical model, whereas the

Table 5 Effect factors of the full factorial design for the system (piroxicam/artichoke waste)

Coefficient	b_0	b_1	b_2	b_3	b_4	b_{12}	b_{13}	b_{23}	b_{14}	b_{24}	b_{34}	b_{123}	b_{124}	b_{134}	b_{234}	b_{1234}
Values	23.2	−1.2	1.8	6.4	−4.6	−1.1	1.1	1.6	−2.9	−3.4	0.4	−1.8	1.7	−0.9	0.6	1.2

Fig. 3 Graphical analysis of the effect of different factors on the piroxicam adsorption onto artichoke waste

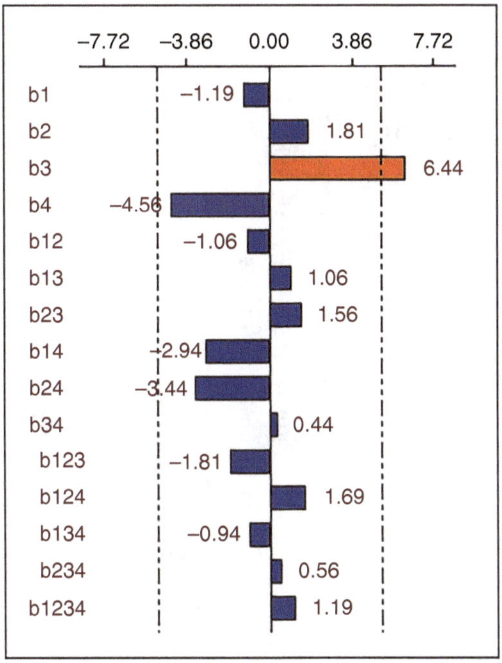

6.37% that exists is attributed to experimental incertitude. The adjusted coefficient of determination was very high $R^2{}_{Adj} = 0.\ 8088 > 0.7$, which implies that the modification efficiency and model choice are strong. In this case and according to the obtained results, the full factorial design can be used to predict and assess the piroxicam adsorption onto artichoke waste.

3.4 *Residue Study of Piroxicam Removal Rate (Henry Plots)*

The statistical inference relating to linear regression is mainly based on the study of the residuals ε, which summarizes the information missing from the mathematical model; it is possible that the model is not correct and that there is an unknown factor of variability that is not integrated into the model. Therefore, it is necessary to analyze the distribution normality of the residues and also the distribution of the residues that must be random around the line [31].

The normal probability graphs are presented in Fig. 5. They show the magnitude of the negative or positive effect of each variable and their interactions on the removal of piroxicam efficiency on artichoke waste. A positive effect value means an increase in the removal efficiency of piroxicam as the factor level increases. On the other hand, a negative effect value means a decrease in the elimination efficiency of piroxicam as the factor level increases. The factors on the right of the centerline

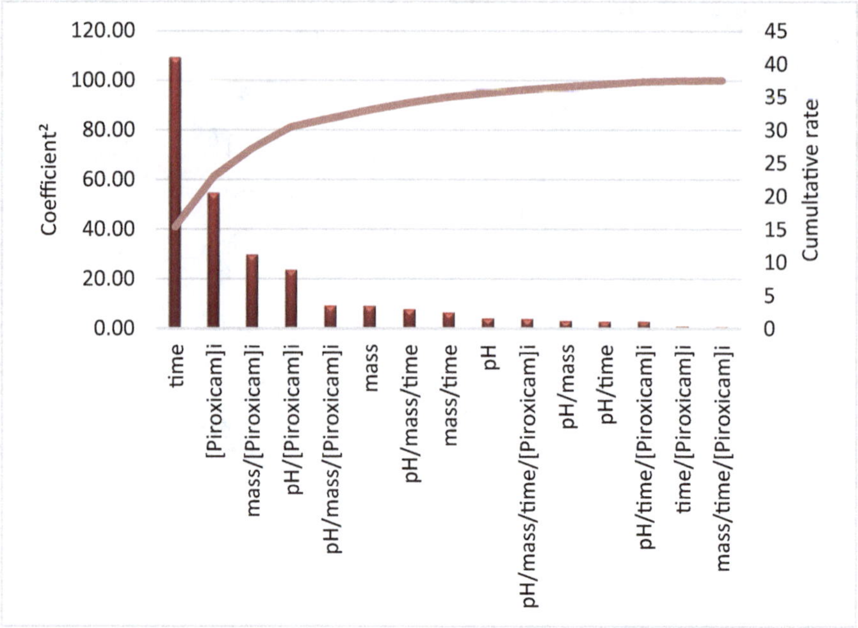

Fig. 4 Graphical Pareto analysis

Table 6 Analysis of variance of full factorial design for piroxicam/artichoke waste system

Variation source	Sum of squares	Degree of freedom	Mean square	Rapport F	Prob F
Model	1.47663×10^3	10	1.47662×10^2	5.2807	0.0200*
Residual	1.39813×10^2	5	2.79625×10^1	$\mathbf{R^2}$	$\mathbf{R^2_{Adj}}$
Total	1.61644×10^3	15		0.936268	0.808803

in Fig. 5b were significant with 1% and have a positive effect (b_3, b_4, b_{14}, and b_{24}) on the elimination efficiency of piroxicam. From the two graphs, we notice that the residuals follow a normal distribution with random distribution around the line $\hat{y}$ (Fig. 6).

4 Conclusions

This work aimed to study the pharmaceutical pollutants removal from wastewater. In this context, piroxicam removal using artichoke is examined in this paper. The obtained results have proved the feasibility of piroxicam adsorption on the studied adsorbent. Also, various parameters were studied, such as the adsorption time, the solution pH, piroxicam's initial concentrations, and the adsorbent mass. Therefore, the obtained results illustrate the following conclusions:

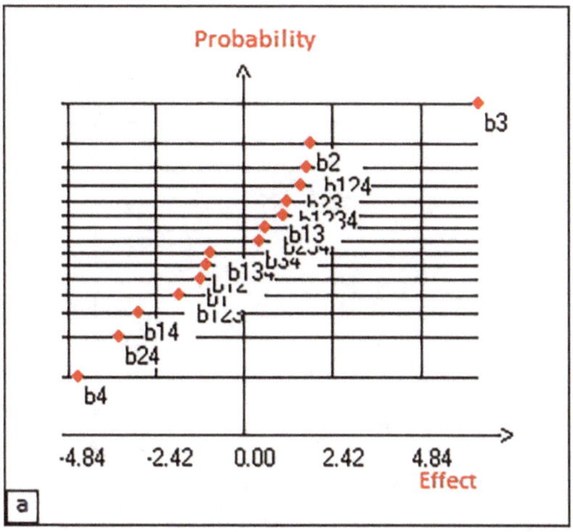

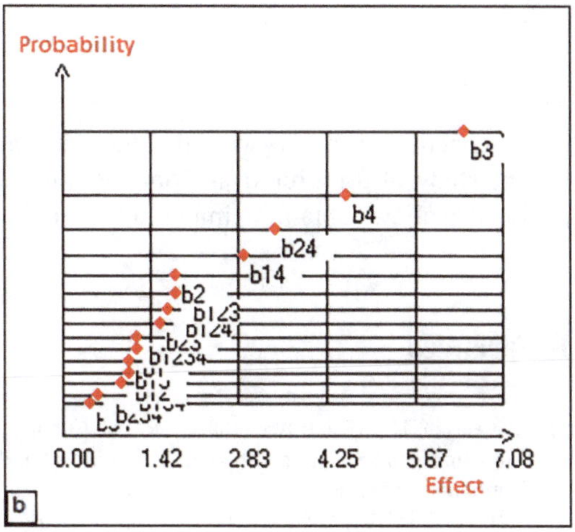

Fig. 5 Normal probability plot (**a**) and half-normal plot (**b**) of the effects

- The significant influence of time and initial concentration on piroxicam removal rate.
- Higher adsorption rates were observed for low initials concentrations of piroxicam.
- Important piroxicam adsorption in acidic medium pH=3.
- The significant influence of mass and the initial piroxicam concentration interaction as well as pH and initial piroxicam concentration interaction on piroxicam removal.

In addition to that, the predicted optimal conditions for the piroxicam removal on artichoke waste are 2 g of artichoke waste, at pH = 3, at a concentration of 10 mg/L, and in a time of 60 min.

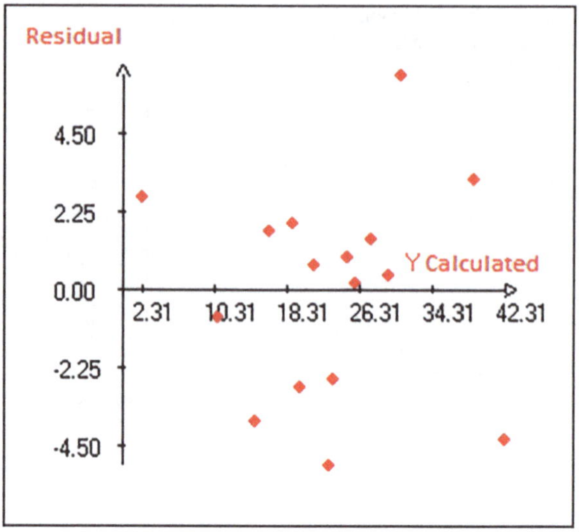

Fig. 6 Residual distribution graph

We conclude from this work that the experimental designs could help in following the study of drug removal from the aqueous medium with fewer experiences. And it also allows obtaining important and efficient results.

References

1. Abu Hasan, H., Sheikh Abdullah, S. R., Al-Attabi, A. W. N., Nash, D. A. H., Anuar, N., Abd. Rahman, N. and Sulistiyaning Titah, H., Removal of Ibuprofen, Ketoprofen, COD and Nitrogen Compounds from Pharmaceutical Wastewater Using Aerobic Suspension-Sequencing Batch Reactor (ASSBR), *Separation and Purification Technology*, vol. **157**, pp. 215–21, January 8, 2016. DOI: https://doi.org/10.1016/j.seppur.2015.11.017
2. Ahmad, A. L., Ismail, S. and Bhatia, S., Optimization of Coagulation–Flocculation Process for Palm Oil Mill Effluent Using Response Surface Methodology, *Environmental Science & Technology*, vol. **39**, no. 8, pp. 2828–34, April 1, 2005. DOI: https://doi.org/10.1021/es0498080
3. Angosto, J. M., Roca, M. J. and Fernández-López, J. A., Removal of Diclofenac in Wastewater Using Biosorption and Advanced Oxidation Techniques: Comparative Results, *Water*, vol. **12**, no. 12, p. 3567, December 19, 2020. DOI: https://doi.org/10.3390/w12123567
4. Azoulay, K., Bencheikh, I., Moufti, A., Dahchour, A., Mabrouki, J. and El Hajjaji, S., Comparative Study between Static and Dynamic Adsorption Efficiency of Dyes by the Mixture of Palm Waste Using the Central Composite Design, *Chemical Data Collections*, vol. **27**, p. 100385, June 1, 2020. DOI: https://doi.org/10.1016/j.cdc.2020.100385
5. Bencheikh, I., Azoulay, K., Mabrouki, J., El Hajjaji, S., Dahchour, A., Moufti, A. and Dhiba, D., The Adsorptive Removal of MB Using Chemically Treated Artichoke Leaves: Parametric, Kinetic, Isotherm and Thermodynamic Study, *Scientific African*, vol. **9**, p. e00509, September 1, 2020. DOI: https://doi.org/10.1016/j.sciaf.2020.e00509
6. Bencheikh, I., Mabrouki, J., Azoulay, K., Moufti, A., and El Hajjaji, S., Predictive Analytics and Optimization of Wastewater Treatment Efficiency Using Statistic Approach, *Big Data and Networks Technologies*, Cham: Springer International Publishing, pp. 310–19, 2020.

7. Bencheikh, I., Azoulay, K., Mabrouki, J., El Hajjaji, S., Moufti, A. and Labjar, N., The Use and the Performance of Chemically Treated Artichoke Leaves for Textile Industrial Effluents Treatment, *Chemical Data Collections*, vol. **31**, p. 100597, February 1, 2021. DOI: https://doi.org/10.1016/j.cdc.2020.100597
8. Ben-Othman, S., Jõudu, I. and Bhat, R., Bioactives from Agri-Food Wastes: Present Insights and Future Challenges, *Molecules*, vol. **25**, no. 3, p. 510, January 2020. DOI: https://doi.org/10.3390/molecules25030510
9. Box, G. E. P., Hunter, W. G. and Hunter, J. S., *Statistics for Experimenters: An Introduction to Design, Data Analysis, and Model Building*, New York: Wiley, pp. 653, 1978.
10. Castrica, M., Rebucci, R., Giromini, C., Tretola, M., Cattaneo, D. and Baldi, A., Total Phenolic Content and Antioxidant Capacity of Agri-Food Waste and by-Products, *Italian Journal of Animal Science*, vol. **18**, no. 1, pp. 336–41, January 2, 2019. DOI: https://doi.org/10.1080/1828051X.2018.1529544
11. Cerino-Córdova, F. J., García-León, A. M., Soto-Regalado, E., Sánchez-González, M. N., Lozano-Ramírez, T., García-Avalos, B. C. and Loredo-Medrano, J. A., Experimental Design for the Optimization of Copper Biosorption from Aqueous Solution by Aspergillus Terreus, *Journal of Environmental Management*, vol. **95**, pp. S77–82, March 1, 2012. DOI: https://doi.org/10.1016/j.jenvman.2011.01.004
12. Chambers, J. M., Cleveland, W. S. and Kleiner, B., *Graphical Methods for Data Analysis*, Wadsworth International Group, pp. 424, 1983.
13. Cheng, S., Zhang, L., Ma, A., Xia, H., Peng, J., Li, C. and Shu, J., Comparison of Activated Carbon and Iron/Cerium Modified Activated Carbon to Remove Methylene Blue from Wastewater, *Journal of Environmental Sciences*, vol. **65**, pp. 92–102, March 1, 2018. DOI: https://doi.org/10.1016/j.jes.2016.12.027
14. Daniel, C., Use of Half-Normal Plots in Interpreting Factorial Two-Level Experiments, *Technometrics*, vol. **1**, no. 4, pp. 311–41, November 1, 1959. DOI: https://doi.org/10.1080/00401706.1959.10489866
15. El Fakir, L., Flayou, M., Dahchour, A., Sebbahi, S., Kifani-Sahban, F. and El Hajjaji, S., Adsorptive Removal of Copper (II) from Aqueous Solutions on Phosphates: Equilibrium, Kinetics, and Thermodynamics, *Desalination and Water Treatment*, vol. **57**, no. 36, pp. 17118–27, August 1, 2016. DOI: https://doi.org/10.1080/19443994.2015.1112840
16. Jaria, G., Silva, C. P., Oliveira, J. A. B. P., Santos, S. M., Gil, M. V., Otero, M., Calisto, V. and Esteves, V. I., Production of Highly Efficient Activated Carbons from Industrial Wastes for the Removal of Pharmaceuticals from Water—A Full Factorial Design, *Journal of Hazardous Materials*, vol. **370**, pp. 212–18, May 15, 2019. DOI: https://doi.org/10.1016/j.jhazmat.2018.02.053
17. Lawson, J., Grimshaw, S. and Burt, J., A Quantitative Method for Identifying Active Contrasts in Unreplicated Factorial Designs Based on the Half-Normal Plot, *Computational Statistics & Data Analysis*, vol. **26**, no. 4, pp. 425–36, February 6, 1998. DOI: https://doi.org/10.1016/S0167-9473(97)00040-6
18. Lenth, R. V., Quick and Easy Analysis of Unreplicated Factorials, *Technometrics*, vol. **31**, no. 4, pp. 469–73, November 1989. DOI: https://doi.org/10.1080/00401706.1989.10488595
19. Luján-Facundo, M. J., Iborra-Clar, M. I., Mendoza-Roca, J. A. and Alcaina-Miranda, M. I., Pharmaceutical Compounds Removal by Adsorption with Commercial and Reused Carbon Coming from a Drinking Water Treatment Plant, *Journal of Cleaner Production*, vol. **238**, p. 117866, November 20, 2019. DOI: https://doi.org/10.1016/j.jclepro.2019.117866
20. Manna, S., Roy, D., Adhikari, B., Thomas, S. and Das, P., Biomass for Water Defluoridation and Current Understanding on Biosorption Mechanisms: A Review, *Environmental Progress & Sustainable Energy*, vol. **37**, December 27, 2017. DOI: https://doi.org/10.1002/ep.12855
21. Marques, S. C. R., Marcuzzo, J. M., Baldan, M. R., Mestre, A. S. and Carvalho, A. P., Pharmaceuticals Removal by Activated Carbons: Role of Morphology on Cyclic Thermal Regeneration, *Chemical Engineering Journal*, vol. **321**, pp. 233–44, August 1, 2017. DOI: https://doi.org/10.1016/j.cej.2017.03.101
22. Mathieu, D., and Nony, J., NEMROD®, SoftwareLPRAI, 1995.

23. Myers, R. H., Montgomery, D. C. and Anderson-Cook, C. M., *Response Surface Methodology: Process and Product Optimization Using Designed Experiments*, John Wiley & Sons, pp. 854, 2016.
24. Nowicki, P., Kazmierczak-Razna, J. and Pietrzak, R., Physicochemical and Adsorption Properties of Carbonaceous Sorbents Prepared by Activation of Tropical Fruit Skins with Potassium Carbonate, *Materials & Design*, vol. **90**, pp. 579–85, January 15, 2016. DOI: https://doi.org/10.1016/j.matdes.2015.11.004
25. Ouzidan, F., Amardo, N., Kouali, M. el and Talbi, M., Optimization of Adsorption of Cationic Dye from Aqueous Solution by Biochar from Artichoke Waste Using Response Surface Methodology, *Mediterranean Journal of Chemistry*, vol. **10**, p. 99, February 3, 2020. DOI: https://doi.org/10.13171/mjc10102002031211fo
26. Ouzidan, F., El Kouali, M., Talbi, M. and Atmani, R., Adsorption of Methylene Blue onto Artichoke Waste, *Oriental Journal of Chemistry*, vol. **31**, no. 4, pp. 2037–41, December 16, 2015.
27. Rakotomalala, R., Pratique de La Régression Linéaire Multiple, p. 190, n.d.
28. Siddeeg, S. M., Amari, A., Tahoon, M. A., Alsaiari, N. S. and Rebah, F. B., Removal of Meloxicam, Piroxicam and Cd+2 by Fe3O4/SiO2/Glycidyl Methacrylate-S-SH Nanocomposite Loaded with Laccase, *Alexandria Engineering Journal*, vol. **59**, no. 2, pp. 905–14, April 1, 2020. DOI: https://doi.org/10.1016/j.aej.2020.03.018
29. Taylor, D. and Senac, T., Human Pharmaceutical Products in the Environment - the "Problem" in Perspective, Chemosphere, 115, pp. 95–99, 2014. https://doi.org/10.1016/j.chemosphere.2014.01.011
30. Ummartyotin, S. and Pechyen, C., Strategies for Development and Implementation of Bio-Based Materials as Effective Renewable Resources of Energy: A Comprehensive Review on Adsorbent Technology, *Renewable and Sustainable Energy Reviews*, vol. **62**, pp. 654–64, September 1, 2016. DOI: https://doi.org/10.1016/j.rser.2016.04.066
31. Vasiliadou, I. A., Molina, R., Martinez, F., Melero, J. A., Stathopoulou, P. M. and Tsiamis, G., Toxicity Assessment of Pharmaceutical Compounds on Mixed Culture from Activated Sludge Using Respirometric Technique: The Role of Microbial Community Structure, *Science of The Total Environment*, vol. **630**, pp. 809–19, July 15, 2018. DOI: https://doi.org/10.1016/j.scitotenv.2018.02.095

Theoretical Study of Thermoelectric Transport Properties of Dicalcium Silicide and Dicalcium Germanide Compounds

A. El Yousfi, H. Bouda, M. L. Ould Ne, A. G. El Hachimi, J. Mabrouki, A. El Kenz, and A. Benyoussef

1 Introduction

Thermoelectricity presents a current research topic, the ability to generate electricity by drawing on heat sources lost in cars, homes, and industries [1, 2]. The discovery of thermoelectric effects is now over 2 centuries, but, until now, the diffusion of devices that exploit this phenomenon remains on a very small scale [3]. The commercialization of thermoelectric devices has never started, usually due to the high cost of such materials [4]. One of the main challenges for the scientific community is to lift this technology gap and develop materials that can greatly improve efficiency and yield by having massive production of cheap, nontoxic, and less restrictive thermoelectric devices for a larger spectrum of applications.

Due to their ability to convert heat into electricity and vice versa, thermoelectric materials attract much interest [5–7]; hence, several theoretical studies based

A. El Yousfi · H. Bouda (✉) · A. El Kenz
Faculty of Sciences, Laboratory of Condensed Matter and Interdisciplinary Sciences (LaMCScI), University of Mohammed V, Rabat, Morocco

M. L. Ould Ne
Institut National de la Recherche Scientifique | INRS · Energy, Materials and Telecommunications Research Centre, Laval, QC, Canada

A. G. El Hachimi
Instituto de Ciencias Físicas, Universidad Nacional Autónoma de México, Cuernavaca, Morelos, Mexico

J. Mabrouki (✉)
Faculty of Sciences, Laboratory of Spectroscopy Molecular Modeling, Materials, Nanomaterials, Water and Environment, University of Mohammed V, Rabat, Morocco

A. Benyoussef
Hassan II Academy of Science and Technology, Rabat, Morocco

M. Azrour et al. (eds.), *IoT and Smart Devices for Sustainable Environment*, EAI/Springer Innovations in Communication and Computing,
https://doi.org/10.1007/978-3-030-90083-0_4

on semiclassical Boltzmann transport theory to calculate thermoelectric transport properties have been achieved [8–13]. The thermoelectric performance of a material depends on its dimensionless figure of merit (ZT); a high value is needed. The figure of merit (ZT) is given as: $\mathrm{ZT} = \frac{\sigma s^2 T}{\kappa_e + \kappa_{ph}}$, where T is the absolute temperature S is Seebeck coefficient. The numerator (σs^2) is also called power factor (PF), which one should maximize. σ is the electrical conductivity. The denominator is the sum of the electronic contribution to thermal conductivity κ_e and the contribution from the phonons (lattice) $\kappa_{ph.}$ A low value of thermal conductivity is needed. Therefore, materials with a high electrical conductivity and Seebeck coefficient have a small thermal conductivity [14]. The materials which are good for thermoelectric applications have a figure of merit ZT value that equals one or more than one [14]. Thermoelectric materials, such as Bi_2Te_3, PbTe, and SiGe, have ZT values around 1 at room temperature [14]. In recent years, tremendous progress has made it possible to synthesize new complex materials with ZT value that can exceed 2 [15]. But, until now, devices that exploit these high ZT have not confirmed the expected efficiency. Most materials currently used in thermoelectric devices have ZTs that do not exceed 1 and that strongly depend on the operating temperature. At room temperature, the Bi_2Te_3 is used for thermoelectric applications since the 1950s [16]. The Bi_2Te_3 material remains the most common material. It has been shown that by combining it with Sb_2Te_3 and Bi_2Se_3, it is possible to regulate carrier concentration and thermal conductivity and maximize the ZT at different temperatures. It is then possible to adjust the material composition for specific applications, such as cooling or power generation [17]. Peak values of ZT for these materials are typically in the range of 0.8–1.1. For average operating temperatures (500–900 K), materials such as PbTe, GeTe, or SnTe are generally used with ZT peaks of about 0.8 [16]. Again, the adjustment of the concentration of the carriers makes it possible to modify the temperature of the peaks of ZT. Recently, for the same temperature range, particular alloys of antimony telluride Sb_2Te_3, germanium GeTe, and silver Ag_2Te (TAGS), with ZT > 1.2, have been used successfully in thermoelectric generators [18]. On the other hand, above 900 K, the n-doped SiGe alloy is conventionally used in devices; the merit figure is much weaker for the same p-doped material. The search for new thermoelectric materials remains open [17]. Silicide compounds such as $BaSi_2$, $FeSi_2$, and Mg_2Si, which are composed of abundant elements in the earth's crust and with economic prices, have attracted attention as environmentally friendly thermoelectric materials [19–21].

On the other hand, first principle calculations play an important role to investigate and predict several novel physical properties that can find several applications such as in optoelectronics and solar cells [22–24], spintronics [25, 26], hydrogen storage [27, 28], thermoelectric, and transport properties [29–31]. In this context, this study explored dicalcium silicide, which contains silicon, the second most abundant element on the earth's crust, making it a quite attractive material. Whereas dicalcium germanide is composed of germanium, which isn't certainly a rare element but is not cheaper as silicon [32, 33]. This work aims to investigate thermoelectric and transport properties of Ca_2X (X = Ge, Si) materials within the first principle calculations implemented on Wien2k code, as well as the Boltztrap package, however, other researches employed VASP code [34, 38]. Thermoelectric parameters are expressed as a function of temperature (300–800 K) and chemical potential. In this study, it is rather difficult to obtain the lattice contribution to the thermal conductivity k_{ph}; we will focus here on the power factor FP $= (\sigma s^2)$ and its individual contributions S and σ.

2 Method of Computation

We used BoltzTraP program to investigate the thermoelectric properties based on the analytical expressions of the electronic bands.

This program calculates semiclassical transport based on the all-electron full potential linear augmented plane wave (FPLAPW) method within the framework of the WIEN2K code [35].

As a punctual value of the bandgap is requested to predict the performance of thermoelectric materials, the corrective approach Tran-Blaha-modified Becke-Johnson (TB-mBJ) exchange potential is used for the electronic band structure calculated [36].

In order to converge the energy eigenvalues, the wave functions in the interstitial regions were expanded in plane waves with cutoff $R_{MT*}K_{max} = 8.0$, where R_{MT} is the muffin-tin radius and K_{max} is the large plane wave vector. The energy convergence criterion was set to be 10^{-5} Ry. For our calculations 5000 k-points in the irreducible of the Brillouin zone was chosen. For chemical potentials range from -1.5 to 1.5 eV the Fermi level being set at 0 eV.

Based on the calculation of the band structure, we have investigated the transport properties of the Cubic structure of Cu_2X (X = Ge, Si), such as electrical conductivities, thermal conductivities, Seebeck coefficients, and power factor, using the standard Boltzmann kinetic transport theory and the rigid band approach [37].

The electrical conductivity, Seebeck coefficient, and thermal conductivity (electronic part) k_0 tensors are functions of temperature (T) and chemical potential (μ) and given by the equations:

$$\alpha_{\alpha\beta}(T;\mu) = \frac{1}{\Omega}\int \sigma_{\alpha\beta}(\varepsilon)\left[-\frac{\partial f_\mu(T;\varepsilon)}{\partial\varepsilon}\right]d\varepsilon \tag{1}$$

$$k^0_{\alpha\beta}(T;\mu) = \frac{1}{e^2T}\int \sigma_{\alpha\beta}(\varepsilon)(\varepsilon-\mu)^2 \times \left[-\frac{\partial f_\mu(T;\varepsilon)}{\partial\varepsilon}\right]d\varepsilon \tag{2}$$

$$S_{\alpha\beta}(T;\mu) = \frac{1}{eT\Omega\sigma_{\alpha\beta}(T;\mu)}\int \sigma_{\alpha\beta}(\varepsilon)(\varepsilon-\mu) \times \left[-\frac{\partial f_\mu(T;\varepsilon)}{\partial\varepsilon}\right]d\varepsilon \tag{3}$$

Where

$$\sigma_{\alpha\beta}(\varepsilon) = \frac{e^2}{N} \sum_{i,k} \tau_i k v_\alpha(i,k)\, v_\beta(i,k)\, \delta\left(\varepsilon - \varepsilon_{i,k}\right) \tag{4}$$

Where α and β, e, and μ are the tensor indicators, electron charge, unit cell volume, and the carrier concentration, respectively. f_μ is the Fermi-Dirac distribution function, and $\boldsymbol{\tau}$ is called the relaxation time, which is the important input for the determination of thermoelectric properties.

3 Results and Discussion

3.1 Energy Band Structure

Figure 1 shows the electronic band structure, respectively, of the cubic system Ca2Ge and Ca2Si, with the lattice constants, respectively, $a = 7.197$ Å (7.016 Å) and direct band gap 0.60 and 0.56 eV at the X point, which is in good agreement with theoretical studies [38, 39].

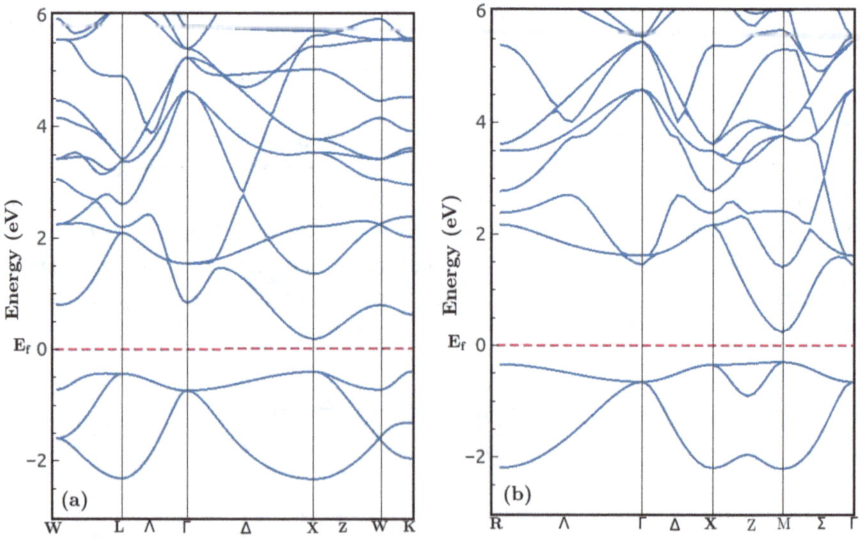

Fig. 1 Band structure of (a) Ca2Ge and (b) Ca2Si.

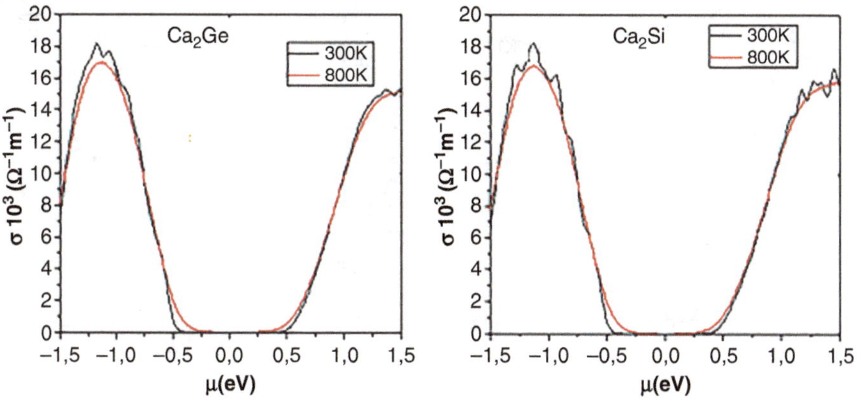

Fig. 2 Electrical conductivity as a function of chemical potential at temperatures for 300 and 800 K

3.2 *Electrical Conductivity (σ)*

The relaxation time τ is required to calculate the electrical conductivity; the output value in BoltzTraP code is σ/τ. However, determining τ is difficult because it depends on both temperature and carrier concentration.

We used an approximate constant relaxation time of 10^{-14} s, which has been used in some recent studies (see Ref. [40]), which causes rigorous limitations in evaluating their PF value.

To minimize the heating of the conductor due to the Joule heating effect, a high electrical conductivity is needed. The electrical conductivity of Ca_2Ge and Ca_2Si are shown in Fig. 2 as a function of chemical potential at 300 and 800 K. The conductivity for Ca_2Ge and Ca_2Si shows the same behavior at 300 and 800 K, at the range of chemical potential (from −0.42 to 0.46 eV), and reaches minimum values in the range. Likewise, the maximum value obtained for the conductivity is $\sigma = 2.59589 \cdot 10^6\ \Omega^{-1} \cdot m^{-1}$ for Cu_2Ge and $\sigma = 2.15706 \cdot 10^6\ \Omega^{-1} \cdot m^{-1}$ for Ca_2Si at $\mu = 1.0$ eV; this value is required for high efficiency **TE** devices.

3.3 *Electronic Thermal Conductivity (κ)*

Thermal conductivity is the proportional coefficient between the gradient temperature and the thermal flux. A material heat is conducted by free electrons and lattice vibrations. The decrease in thermal conductivity increases the thermoelectric performance of materials.

The electronic contribution of thermal conductivity Of Ca_2Ge and Ca_2Si is given against chemical potential at both $T = 300$ K and 800 K as shown in Fig. 3.

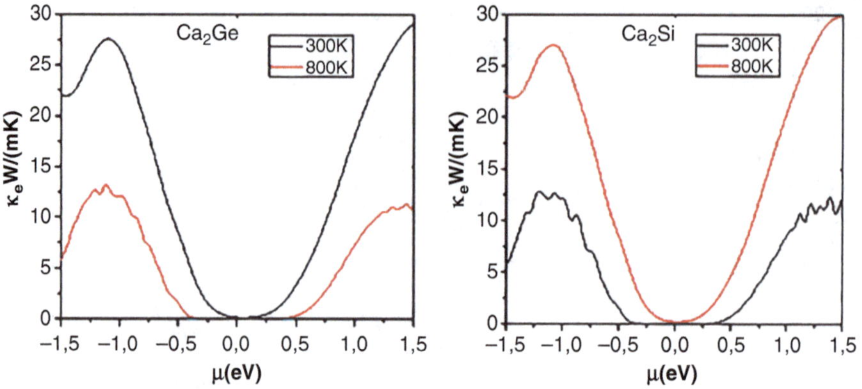

Fig. 3 Thermal conductivity in function of chemical potential at temperatures for 300 and 800 K

The thermal conductivity of Ca_2Ge is at their minimum values at the range of chemical potential (−0.42 and 0.04 eV) for $T = 800$ K, while Ca_2Si compound at the range of the minimum thermal conductivity is (−0.40 and 0.36 eV) for $T = 300$ K, and minimum value for the thermal conductivity for both materials is $\kappa_e = 4.0278.10^{-3}$ W/m K. It is clear that the Ca_2Ge material gives a good thermoelectric response at the temperature 800 K , however, this compound shows a reduced thermoelectric value at 300 K, this is due to the large value of the thermal conductivity as depicted in Fig. 3.

3.4 Seebeck Coefficient

The Seebeck coefficient is one of the major thermoelectric (TE) properties. The materials with higher Seebeck coefficient will have high ZT (figure of merit) that means is a good candidate for thermoelectric applications.

The Seebeck coefficient is related to electron carrier (charge and heat). The Seebeck coefficient of a material is the measurement of the magnitude of an induced voltage in response to the temperature gradient across that material.

Figure 4 presents Seebeck coefficient versus chemical potentials for Ca_2Ge and Ca_2Si compounds at room temperature 300 and 800 K; there is a decrease of Seebeck coefficient with an increase of temperature in both cases. For the chemical potential μ equal to −0.005 eV, Seebeck coefficient is positive as the intrinsic charge carriers are holes. Ca_2Ge and Ca_2Si materials displayed an interesting thermoelectric property at room temperature (300 K) such as a very high Seebeck coefficient, respectively, 1700 μV/K (μ = -0.05 eV) and 1550 μV/K (μ = −0.09 eV). From the Seebeck coefficient, it's clear that the materials are a p-type semiconductor.

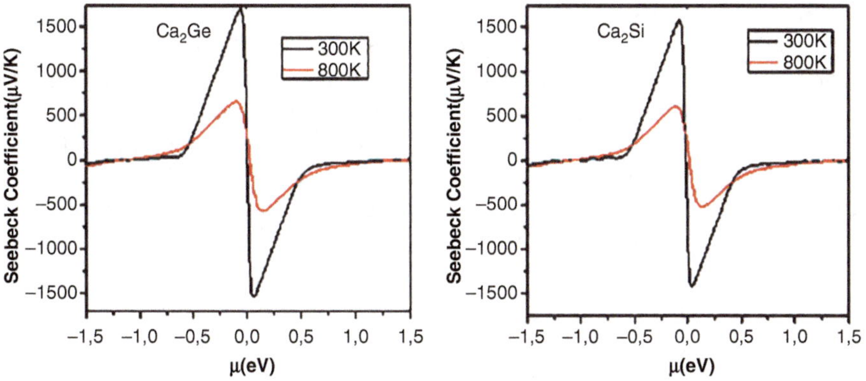

Fig. 4 Seebeck coefficient in function of chemical potential at temperatures for 300 and 800 K

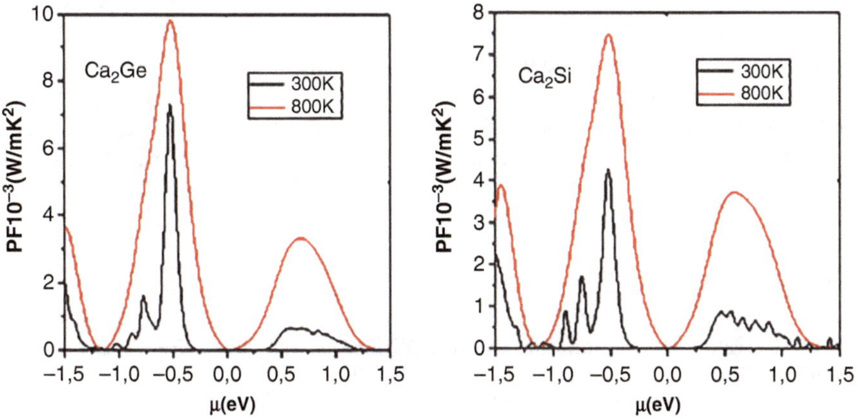

Fig. 5 Power factor (PF) as a function of chemical potential at 300 and 800 K

3.5 Power Factor

The thermoelectric performance of a material is characterized by the power factor. The thermoelectric efficiency of material requires an enhanced power factor and reduced thermal conductivity. By coupling these two expressions S and σ, we found the power factor value given by: PF = $(S^2\sigma)$. The result of PF is represented in Fig. 5; it exhibits similar dependence on chemical potential for both Ca_2Ge and Ca2Si at different temperatures.

The power factor initially increases and then decreases with increasing chemical potential. It exhibits similar dependence on chemical potential for both Ca_2Ge and Ca_2Si at different temperatures. However, Ca2Ge and Ca2Si possess significantly low power factor for $T = 300$ K, which is due to the low electrical conductivity at the range of chemical potential (from −0.19 to 0.44 eV).

The highest value of PF of Ca_2Ge is 9.7×10^{-3} W/mK^2 (at 800 K) at $\mu = -0.52$ eV. On the other hand, for Ca2Si, the optimal value of PF is 7.5×10^{-3} W/mK^2 ($\mu = -0.29$ eV) at 800 K; the result of PF of $CaSi_2$ is according to the theoretical value reported [14]. At room temperature, the high values of PF_s are, respectively, (7.1×10^{-3}) 4.5×10^{-3} W/mK^2 ($\mu = -0.61$ eV) for Ca_2Ge and Ca_2Si. The high values of the PF for both cases, Ca_2Ge and Ca_2Si, at room temperature are comparable with the corresponding ones in forefront thermoelectric materials such as SnSe [41] and $Sb2Se_3$ [42].

4 Conclusions

Using semiclassical Boltzmann transport theory, in combination with first principle calculations, we have studied the thermoelectric properties of bulk Ca_2X (X = Ge, Si) at $T = 300$ K and 800 K. In particular, we computed the Seebeck coefficient, the electrical conductivity, and the electronic thermal conductivity. Our results show that a large power factor is possible for Ca_2Ge (9.7×10^{-3}) 7.1×10^{-3} W/mK^2 at 300 (800) K and for Ca_2Si is (7.5 x10^{-3}) 4.5 x10^{-3} W/K^2. Our findings suggest that Ca_2X (X = Ge,Si) are a promising materials for thermoelectric applications.

References

1. D. S. Dalafave, « Thermoelectric properties of Re6GaxSeyTe15−y (0≤x≤2; 0≤y≤7.5) », Mater. Chem. Phys., vol. 119, no 1 2, p. 195 200, 2010.
2. Bubnova, Olga, Zia Ullah Khan, Abdellah Malti, Slawomir Braun, Mats Fahlman, Magnus Berggren, and Xavier Crispin. "Optimization of the thermoelectric figure of merit in the conducting polymer poly (3, 4-ethylenedioxythiophene)." Nature materials 10, no. 6 (2011): 429–433.
3. Y. Pei, X. Shi, A. LaLonde, H. Wang, L. Chen, et G. J. Snyder, « Convergence of electronic bands for high performance bulk thermoelectrics », Nature, vol. 473, no 7345, p. 66 69, mai 2011.
4. Rowe, David Micheal, Thermoelectrics handbook: Macro to nano. 2005.
5. Baranowski, L. L. Snyder, G. J. & Toberer, E. S. Concentrated thermoelectric generators. Energy Environ. Sci. 5, 9055 (2012).
6. Bell, L. E. Cooling, heating, generating power, and recovering waste heat with thermoelectric systems. Science 321, 1457 (2008). 3.
7. Snyder, G. J., Toberer, E. S., Khanna, R. & Seifert, W. Improved thermoelectric cooling based on the Thomson effect. Phys. Rev. B 86, 045202 (2012).
8. Singh, S. & Gupta, D.C. J Supercond Nov Magn (2019) 32: 2009. doi:https://doi.org/10.1007/s10948-018-4907-1
9. Singh, S., Bhat, T.M. & Gupta, D.C. J Supercond Nov Magn (2019) 32: 2051. doi:https://doi.org/10.1007/s10948-018-4915-1
10. Khandy, Saveer Ahmad, and Dinesh C. Gupta. "Magneto-electronic, mechanical, thermoelectric and thermodynamic properties of ductile perovskite Ba2SmNbO6." Materials Chemistry and Physics 239 (2020): 121983.

11. Bhandari, U., Bamba, C.O., Malozovsky, Y. and Bagayoko, D., 2018. Predictions of Electronic, Transport, and Structural Properties of Magnesium Sulfide (MgS) in the Rocksalt Structure. Journal of Modern Physics, 9(9), pp. 1773–1784.
12. Zhao, G.L., Gao, F. and Bagayoko, D., 2018. Reliable density functional calculations for the electronic structure of thermoelectric material ZnSb. AIP Advances, 8(10), p. 105211.
13. Bohara, B., Franklin, L., Malazovsky, Y. and Bagayoko, D., Ab-initio Calculations of Electronic and Transport Properties of Calcium Fluoride (CaF2)]
14. H. A. Rahnamaye Aliabad et M. Kheirabadi, « Thermoelectricity and superconductivity in pure and doped Bi2Te3 with Se », Phys. B Condens. Matter, vol. 433, p. 157 164, janv. 2014.
15. Snyder, G. J. and E. S. Toberer, « Complex thermoelectric materials », Nat. Mater., vol. VOL 7, p. 105, févr. 2008.
16. Heikes, Robert R., and Roland W. Ure., « Thermoelectricity: science and engineering », Intersci. Publ., 1961.
17. V. L. Kuznetsov, L. A. Kuznetsova, A. E. Kaliazin, et D. M. Rowe, « High performance functionally graded and segmented Bi2Te3-based materials for thermoelectric power generation », J. Mater. Sci., vol. 37, no 14, p. 2893–2897, 2002.
18. Skrabek, E. and D. Trimmer, CRC Handbook of Thermoelectrics. 1995.
19. K. Hashimoto, K. Kurosaki, H. Muta, et S. Yamanaka, « Thermoelectric properties of La-doped BaSi2 », Mater. Trans., vol. 49, no 8, p. 1737–1740, 2008.
20. Li Han, Tang Xin-Feng, Cao Wei Qiang and Zhang Qing Jie, « Condensed Matter: Electronic Structure, Electrical, Magnetic, And Optical Properties: Quick preparation and thermal transport properties of nanostructured β-FeSi2 bulk material », Chin Phys B, vol. 18, no 287 292, 2009.
21. Q. Zhang, X. B. Zhao, H. Yin, et T. J. Zhu, « Thermoelectric performance of Mg2−xCaxSi compounds », J. Alloys Compd., vol. 464, no 1 2, p. 9 12, Sept. 2008.
22. El Yousfi, A., Bouda, H., El Hachimi, A.G., Arshad, M.A., El Kenz, A. and Benyoussef, A., 2021. Enhanced optical absorption of rutile TiO 2 through (Sm, C) codoping: a first-principles study. Optical and Quantum Electronics, 53(2), pp. 1–12.
23. Rai, D.P., Sandeep, Shankar, A., Khenata, R., Reshak, A.H., Ekuma, C.E., Thapa, R.K. and Ke, S.H., 2017. Electronic, optical, and thermoelectric properties of Fe2+ x V1− x Al. AIP advances, 7(4), p. 045118.
24. Chettri, S., Rai, D.P., Shankar, A., Khenata, R., Ghimire, M.P., Thapa, R.K. and Bin Omran, S., 2016. GGA+ U and mBJ+ U study of the optoelectronic, magnetic and thermoelectric properties of the SmAlO3 compound with spin–orbit coupling. International Journal of Modern Physics B, 30(12), p. 1650078.
25. Rai, D.P., Shankar, A., Sakhya, A.P., Sinha, T.P., Khenata, R., Ghimire, M.P. and Thapa, R.K., 2016. Electronic and magnetic properties of X2YZ and XYZ Heusler compounds: a comparative study of density functional theory with different exchange-correlation potentials. Materials Research Express, 3(7), p. 075022.
26. Bouda, H., Bahlagui, T., Masrour, R., Bahmad, L. and Benyoussef, A., 2019. Unexpected magnetic behavior of Ga doped CuFe 1-x Ga x O 2 delafossite, x= 0.04: First principle calculation and Monte Carlo simulation. The European Physical Journal Plus, 134(10), pp. 1–9
27. Raza, H.H., Murtaza, G. and Khalil, R.M.A., 2019. Optoelectronic and thermal properties of LiXH3 (X= Ba, Sr and Cs) for hydrogen storage materials: a first principle study. Solid State Communications, 299, p. 113659.
28. Reshak, A.H., 2013. MgH2 and LiH metal hydrides crystals as novel hydrogen storage material: Electronic structure and optical properties. International journal of hydrogen energy, 38(27), pp. 11946–11954.
29. Rai, D.P., Shankar, A., Ghimire, M.P., Khenata, R. and Thapa, R.K., 2015. Study of the enhanced electronic and thermoelectric (TE) properties of ZrxHf1− x− yTayNiSn: a first principles study. RSC Advances, 5(115), pp. 95353–95359.

30. Yousuf, S. and Gupta, D.C., 2018. Chemical potential evaluation of thermoelectric and mechanical properties of Zr 2 CoZ (Z= Si, Ge) Heusler alloys. Journal of Electronic Materials, 47(4), pp. 2468–2478.
31. Yousuf, S. and Gupta, D.C., 2019. Thermoelectric response of ZrNiSn and ZrNiPb Half-Heuslers: applicability of semi-classical Boltzmann transport theory. Results in Physics, 12, pp. 1382–1386.
32. V. K. Zaitsev, M. I. Fedorov, E. A. Gurieva, I. S. Eremin, P. P. Konstantinov, A. Y. Samunin and M. V. Vedernikov, « Highly effective Mg2Si1-xSnx thermoelectrics », Phys. Rev. B, vol. 74, p. 045207, juill. 2006.
33. http://www.chemicool.com/elements/
34. R. Xiong et al., « Structural stability and thermoelectric property optimization of Ca 2 Si », RSC Adv., vol. 7, no 15, p. 8936 8943, 2017.
35. P. Blaha, K. Schwarz, G. K. H. Madsen, D. Kvasnicka, et J. Luitz, « wien2k », Augment. Plane Wave Local Orbitals Program Calc. Cryst. Prop., 2001.
36. F. Tran et P. Blaha, « Accurate Band Gaps of Semiconductors and Insulators with a Semilocal Exchange-Correlation Potential », Phys. Rev. Lett., vol. 102, no 22, juin 2009.
37. G. K. Madsen et D. J. Singh, « BoltzTraP. A code for calculating band-structure dependent quantities », Comput. Phys. Commun., vol. 175, no 1, p. 67–71, 2006.
38. D. B. Migas, L. Miglio, V. L. Shaposhnikov, et V. E. Borisenko, « Comparative study of structural, electronic and optical properties of Ca 2 Si, Ca 2 Ge, Ca 2 Sn, and Ca 2 Pb », Phys. Rev. B, vol. 67, no 20, mai 2003.
39. Y. Zhiwen et L. Tingju, « First-principle studies the power factor of Ca-X(X=Si,Ge,Sn,Pb) intermatallic compounds », J. Solid State Chem., vol. 183, no 1, p. 136 143, janv. 2010.
40. A. Phillip B, P. Warren E, et K. Henry, « Anisotropic normal-state transport properties predicted and analyzed for high-Tc oxide superconductors », Phys. Rev. B, p. 7482–7490, 1988.
41. Zhao, L.-D.; Tan, G.; Hao, S.; He, J.; Pei, Y.; Chi, H.; Wang, H.; Gong, S.; Xu, H.; et Dravid, V. P., « Ultrahigh power factor and thermoelectric performance in hole-doped single-crystal SnSe », Science, vol. 351, p. 141 144, 2016.
42. C. Donghyeuk et al., « Diameter-Controlled and Surface-Modified Sb2Se3 Nanowires and Their Photodetector Performance », Nature, vol. 4, p. 6714, 2014.

Modeling and Design of Water Treatment Processes by Biosorption Method Using JMP® 11 Software

Karima Azoulay, Imane Bencheikh, Nora Samghouli, Jamal Mabrouki, Ahmed Moufti, and Souad El Hajjaji

1 Introduction

Pollution caused by industrial wastewaters has become a standard problem for several countries [1]. The effluents generated from dye manufacturing and industrial activities can be very colorful and difficult to treat [2].

The removal of dyes from colored effluents, particularly from textile industries is one among the main environmental concern these years [3, 4]. Several methods are employed to get rid of dyes from the wastewater like physical and chemical processes. Among the varied commercial processes, color removal using activated carbon has an important potential [5, 6]. However, due to the high cost of activated charcoal, research is conducted on the utilization of low-cost materials as adsorbents. Though several studies using low-cost adsorbents have successfully applied for removing different types of dyes from solution, only a few of them might be employed effectively [7, 8]. Though several studies using cheaper material as adsorbents have successfully applied for removing different sorts of dyes from solution, only a few of them might be employed effectively [9, 10]. There has

K. Azoulay (✉) · I. Bencheikh · N. Samghouli · J. Mabrouki · S. El Hajjaji (✉)
Faculty of Sciences, Laboratory of Spectroscopy, Molecular, Modelling, Materials, Nanomaterials, Water and Environment, (LS3MN2E), Department of Chemistry, Mohammed V University in Rabat, Rabat, Morocco
e-mail: karima.azoulay@um5r.ac.ma

A. Moufti
Equipe Science de l'Environnement et Matériaux Appliqué (SEMA), Université Sultan Moulay Sliman, Khouribga, Morocco

Environmental Sciences and Applied Materials (ESAM) Team, Khouribga Ply-disciplinary Faculty (FPK), Khouribga, Morocco

Regional Center for Careers in Education and Training, Casablanca-Settat, Morocco

M. Azrour et al. (eds.), *IoT and Smart Devices for Sustainable Environment*, EAI/Springer Innovations in Communication and Computing,
https://doi.org/10.1007/978-3-030-90083-0_5

recently been great interest in biosorbents (low cost) for the removal of dyes from wastewater, which annually produce around 100,000 tonnes of palm waste. These wastes are often burned, used as fuel for cooking and heating, or disposed of at random. The current research focuses on the adsorption of methylene blue (MB) and methyl orange (MO) dyes onto raw palm waste in an aqueous solution. The impact of various influencing parameters on the efficiency of dye removal onto raw palm waste has been elaborated. This step consists in determining the mathematical function, making it possible to describe the phenomenon studied with the influencing factors. An approach based on the use of response surface methodology (RSM) [11, 12] has been used to address optimization problems. This methodology is simple, efficient, and easy to apply. It allows to extract information by a minimum of experimental tests (the experiment is expensive in terms of time and resources) in order to characterize a process as precisely as possible. RSM is the second part of the design of experiment method, which helps to better understand and optimize the reaction parameters in order to achieve desirable responses. This response surface behavior has been explained by a complete second-degree polynomial model, which allows the estimation of linear, quadratic, and interaction effects. For this, an experimental design software, JMP version 11, offering a multitude of statistical and graphical possibilities was used to design the experiments, identify the influencing factors, and optimize the models.

2 Materials and Methods

2.1 Materials

2.1.1 Adsorbents

The material used in the present work was the raw date seed of Drâa-Tafilalet, Errachidia, Morocco. The raw date seeds were washed with hot distilled water to remove the impurities and surface-adhered particles and then dried at 80 °C until the mass be stabilised. The dried samples were milled and then screened. The particle diameters in the range of 0.250 and 0.315 mm was selected for the experiments.

2.1.2 Adsorbates

Methylene blue (MB) is an organic compound and a cationic dye with the chemical formula C16H18N3SCl (Fig. 1). Its molar weight is 319.85 g/mol [13].

Methyl orange (MO) is an organic compound and an anionic dye with the chemical formula C14H14N3NaO3S (Fig. 2). Its molar weight is 327.33 g/mol [14].

The stock solution of 20 mg/L MB and MO was prepared by dissolving appropriate amounts of MB and MO in distilled water. The MB and MO stock solution could be diluted to obtain solutions of different concentrations.

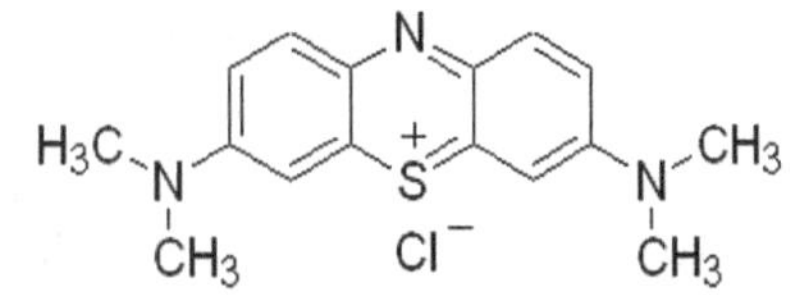

Fig. 1 Chemical structures of methylene blue

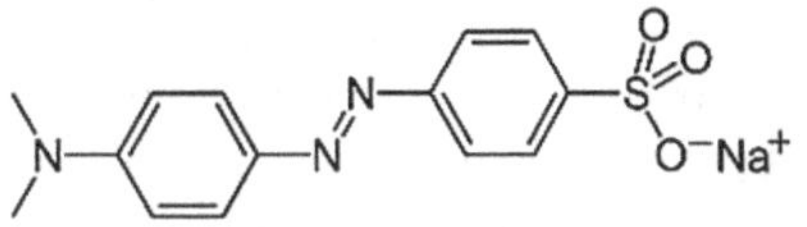

Fig. 2 Chemical structures of methyl orange

3 Methods

3.1 Experimental Method

An amount of adsorbents was transferred into 100 mL of MB and MO solutions at different concentrations (5–20 ppm). The mixture solutions were stirred at room temperature and withdrawn at predetermined time intervals. At regular intervals, the samples were separated by the centrifuge, and the final concentration was measured by UV-visible spectrophotometer, with the maximum absorbance wavelength being 665 for MB and 465 for MO. In order to find the optimal conditions, the studied response is the retention rate ($R\%$). It was calculated as follows:

$$\%R = 100 \times \frac{(C_0 - C_e)}{C_0} \tag{1}$$

Where:

C_0: Initial concentrations of dyes (at $t = 0$) in mg L^{-1}
C_e: Equilibrium concentrations of dyes (at $t = t_e$) in mg L^{-1}

4 Design of Experiments

Response surface methodology (RSM) is a technique for determining variations in response to significant influencing factors. This method makes it possible to determine an approximation relationship between the input variables and the output variables [15–17]. The response surface methodology goes through three stages: construction of the experimental design, response modeling, and graphic representations. The design used in an RSM study is quadratic design such as the central composite design units (Box-Wilson) or Box-Behnken. Also, RSM can be used to estimate the effects of individual parameters, the interaction of variables, and the optimum conditions for responses [18, 19].

5 Selected Parameters

In this study, we chose the three factors, which are: the initial potential of hydrogen (pH) of the aqueous medium, the dose of raw palm waste, and the initial concentration of the dye [20]. These factors are those that have a priori direct influence on the biosorption of the dye, i.e., on the removal efficiency and on the biosorption capacity. For this optimization study, it is obligatory to use a second degree polynomial model. Among the designs that allow the use of a second degree polynomial model, we have chosen the central composite design unit (Box-Wilson), which allows optimal qualities in terms of the forecast of the response calculated at all points of the domain. The optimum conditions for these three parameters have been determined using statistical experimental design. For that, the limit of variation of these factors must be chosen, which determines the experimental domain; each factor requires two levels (high levels (+1) and low levels (−1)). These parameters are grouped in the following table:

Seventeen trials were necessary (2^3+ 2*3 + 3) for the central composite design (Table 1). All experiments were performed in random order, and the calculations were performed using the JMP® 11 (Discovering JMP. Cary, NC: SAS Institute Inc.) software [21]. The functional conditions of the tests by the central composite design are grouped in Table 2:

General model polynomial equation for the second-order degree is written as follows:

$$y = \beta_0 + \sum_{i=1}^{3} \beta_i x_i + \sum_{i=1}^{3} \sum_{j=1}^{3} \beta_{ij} x_i x_j + \sum_{i=1}^{3} \beta_{ii} x_i 2 + \varepsilon \tag{2}$$

Where:

y is the predicted response (lead removal capacity)
x_i and x_j are the independent variables
β_0, β_i, β_{ii}, and β_{ij} are the model constant, the linear coefficients, the quadratic coefficients, and the cross-product coefficients, respectively ε represents the random error.

Table 1 The levels and ranges of each variable (factors)

			Levels	
Variables (factors)	Code	Units	−1	+1
pH	X_1	–	2	11
Mass	X_2	mg	25	100
[Dyes]	X_3	mg/L	5	20

Table 2 Experiment matrix of the CCD

	Experimental value		
Run	pH	Mass (mg)	[dye] (mg/L)
1	0.7	75	12.5
2	2	50	5
3	2	50	20
4	2	100	5
5	2	100	20
6	6.5	42.8	12.5
7	6.5	75	2.85
8	6.5	75	12.5
9	6.5	75	12.5
10	6.5	75	12.5
11	6.5	75	22.15
12	6.5	107.18	12.5
13	11	50	5
14	11	50	20
15	11	100	5
16	11	100	20
17	12.3	75	12.5

6 Coefficient of Determination R^2

Analysis of variance allows calculating statistical parameter: the R^2 or R squared. This statistic is the ratio of the sum of the squares of the calculated responses (corrected for the mean) to the sum of the squares of the measured responses (corrected for the mean) [22]:

$$R^2 = \frac{\sum_{i=1}^{n} \left(\hat{y}\mathrm{i} - \overline{y}\right)^2}{\sum_{i=1}^{n} (y\mathrm{i} - \overline{y})^2} \tag{3}$$

$R^2 = 1$ indicates a perfect fit; on the other hand, an R^2, which is equal to 0, indicates the absence of relationship between the dependent variable and the explanatory variable. However, in the context of multiple regressions, this poses the problem of model parameterization [23]. The more explanatory variables are added, the more the R^2 increases. To avoid this phenomenon, we calculate the adjusted coefficient of determination. The quality of the model will therefore be better if the adjusted R^2 will be close to 1.

7 Analysis of Variance

Analysis of variance (ANOVA) is a necessary tool to desire the meaning of a parameter or a mathematical model. The principle of the analysis of variance is based on the calculation of the total difference between the different measurements, y_i, of the design of experiments and the average of these measurements [24].

8 Study of the Interactions by the Response Surface

Interaction profiles of the three studied factors was used to understand the interactions between variables and determine the optimum level of every variable for the discoloration efficiency [25].

9 Optimization

The goal of this optimization is to meet an objective, which is in our case the biosorption of dyes (MB and MO). It therefore consists in finding all of the values of the operating variables (factors) that lead to the desired response, based on economic constraints. After the analysis of the system and modeling of the response according to various factors, the optimum can be located by the method of plotting isoresponse curves [26].

10 Results and Discussion

10.1 Experimental Design

Table 2 shows the experimental design matrices with the two response values obtained from the experimental work. Midpoint runs 8, 9, and 10 were performed to determine experimental error and data reproducibility. JMP software was used to design the assays. Table 3 represents the experimental tests with all the input parameters, which were carried out.

11 Coefficient of Determination

The adsorption determination coefficient of R^2_{MB} and R^2_{MO}, having values 0.98 and 0.99, respectively, is near to 1. These values give a good compatibility between the

Table 3 Experimental design matrix for raw date seeds

	Experimental value	Retention rate *R*%			
Run	Ph	Mass (mg)	[Dye] (mg/L)	MB	MO
1	0.7	75	12.5	62.36	38.67
2	2	50	5	55.49	47.32
3	2	50	20	33.61	29.64
4	2	100	5	80.07	58.10
5	2	100	20	47.47	34.39
6	6.5	42.8	12.5	98.70	15.36
7	6.5	75	2.85	85.83	46.76
8	6.5	75	12.5	90.62	21.31
9	6.5	75	12.5	98.64	15.56
10	6.5	75	12.5	98.52	17.62
11	6.5	75	22.15	98.80	1.75
12	6.5	107.18	12.5	99.69	17.10
13	11	50	5	80.13	26.00
14	11	50	20	78.76	13.34
15	11	100	5	99.85	24.21
16	11	100	20	83.58	18.60
17	12.3	75	12.5	83.78	14.43

values, experimental and planned, of the adapted model. Figures. 3a and 3b confirm that the model is relatively well adjusted. Therefore, there is a good correlation between the measured values and the calculated values.

12 Analysis of Variance (ANOVA)

The results gated from the analysis of the variance of MB and MO are presented in Table 4.

Table 4 shows that the polynomial model (central composite design) is significant at the confidence level of 0.05, given that *p*-values are less than 0.05 for MB and MO. The interaction between the pH of the solution and the concentration of methylene blue and the double interaction of the pH of two dyes and the concentration of MO were statistically significant on the removal of methylene blue and methylene orange, with a *p*-value of 0.0033, 0.0448, 0.0036, and 0.0013, respectively. On the other hand, the effects of mass, the concentration of interaction dyes between pH and mass, the interaction between mass and the concentration of OM, the double interaction of mass, and the double interaction of the concentration of dyes were not important [27].

Fig. 3 Experimental and predicted response for (**a**) MO and (**b**) MB removals

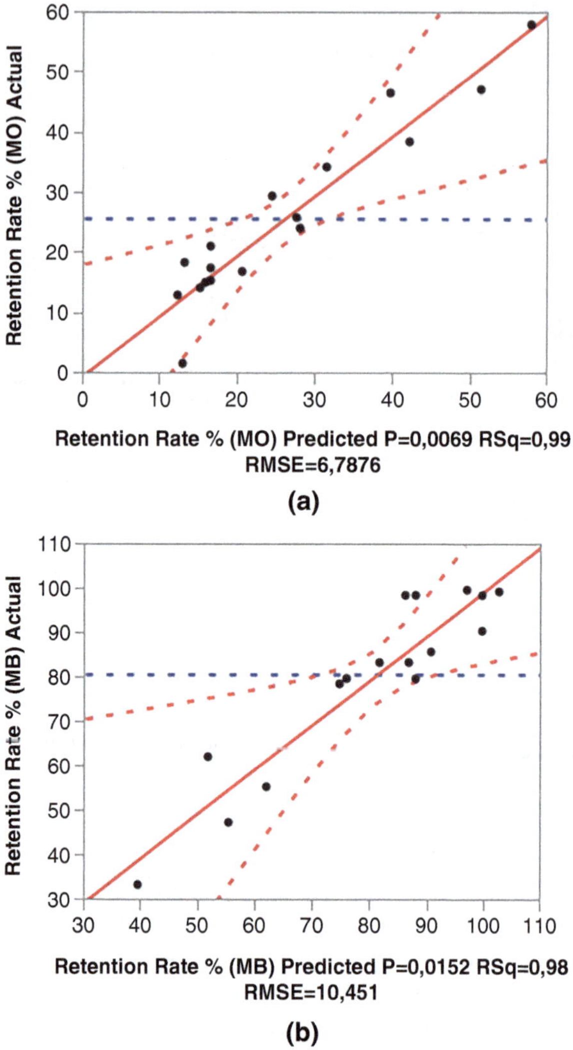

13 Development of Regression Model Equation

According to the sum of the squares of the sequential model selected on the basis of the highest order polynomials, the complementary terms were significant. For the two responses, methylene blue (MB) and methyl orange (MO), the quadratic models were selected as suggested by the JMP software. The final empirical models in terms of coded factors (parameters) for the removal of MB and MO (R) are represented by Eqs. (4) and (5), respectively:

Table 4 Quadratic polynomial model of MB and MO

	DF	Estimate	Sum of squares	*t* ratio	*P*-value					
Term	MB	OM	MB	OM	MB	OM	MB	OM	MB	OM
model	9	9	99.58	16.43	5694.29	3170.71	5.79	7.64	<.0001*	0.0019*
$X_{1,}$ pH	1	1	13.5	1.87	2075.77	1240.91	4.36	−5.19	0.003*	0.0013*
X_2, mass	1	1	5.6	−10.39	365.04	39.86	1.83	0.93	0.1103	0.3830
$X_{3,}$ [dye]	1	1	−1.7	−1.50	35.90	1222.31	−0.57	−5.15	0.5846	0.0013*
$X_1 * X_2$	1	1	−1.7	2.88	24.14	18.20	−0.47	−0.63	0.6525	0.5496
$X_1 * X_3$	1	1	9.01	0.12	650.31	66.73	2.44	1.20	0.044*	0.2679
$X_2 * X_3$	1	1	0.5	7.27	2.17	0.13	0.14	0.05	0.8915	0.9588
$X_1 * X_1$	1	1	−18.4	1.04	2010.42	312.01	−4.29	2.60	0.0036*	0.0353*
$X_2 * X_2$	1	1	−2.7	5.89	42.98	6.46	−0.63	0.37	0.5504	0.7190
$X_3 * X_3$	1	1	−6.8	−10.47	276.75	204.57	−1.59	2.11	0.1555	0.0731
Lack of fit	5	5	–	–	722.39	305.51	6.84	7.19	–	–
Error	7	7	–	–	764.63	322.5	–	–	–	–
C.total	16	16	–	–	6458.92	3493.21	–	–	–	–

* means that the value is below 0.05

$$R\,(\mathrm{BM}) = 99.58 + 13.54 \times 1 - 18.46 \times 12 + 9.015 \times 13 \tag{4}$$

$$R\,(\mathrm{OM}) = 16.43 - 10.47 \times 1 - 10.39 \times 3 + 7.27 \times 12 \tag{5}$$

14 Analysis of Interactions

According to Fig. 4, the interaction between the pH and the concentration of the MB means that the effect of pH of the solution of the MB is not the same on the response when we change the concentration of the solution, whereas for the MO, there is no important interaction. Figure 4 represents the interactions of the parameters for both dyes.

15 Optimization

15.1 Prediction Profiler

The aim of parameter optimization is to find the optimal setting of the three parameters that leads to maximizing the removal efficiency of 99.5% BM dye and 54.72% OM dye. In other words, we will seek the optimal operating conditions that lead to this performance while minimizing the mass of raw palm waste and maximizing the concentration of dyes. In order to determine the exact optimal condition, we apply the function (desirability); it allows giving an exact optimal adjustment, which varies between 0 and 1 [28]. Moreover, the attribution of the value of 0 is done when the factors give an answer undesirable (unacceptable) and the value of 1 in the case where the response represents a maximum performance desired for the factors. Figure 5 indicates that achieving these yields is possible with a desirability of around 99.22% and 52.38% for BM and MO, respectively. Therefore, the operating conditions of soution's pH ($pH_{MB} = 8.1$; $pH_{MO} = 2$), adsorbent mass ($m_{MB} = 64.62$ mg; $m_{MO} = 50$ mg), and adsorbate concentration ([BM] = 13.54 mg/L; [OM] = 5 mg/L).

16 Three-Dimensional Response Surface (3-D)

The objective of this optimization is to find the set of values of the factors leading to the desired response, which is the optimum efficiency of the removal of dyes from the raw palm waste. The use of contour graphs (3-D) is considered to be an easy technique for the optimization of the treatment ratio and even for identification between the studied parameters. It makes it possible in particular to trace and

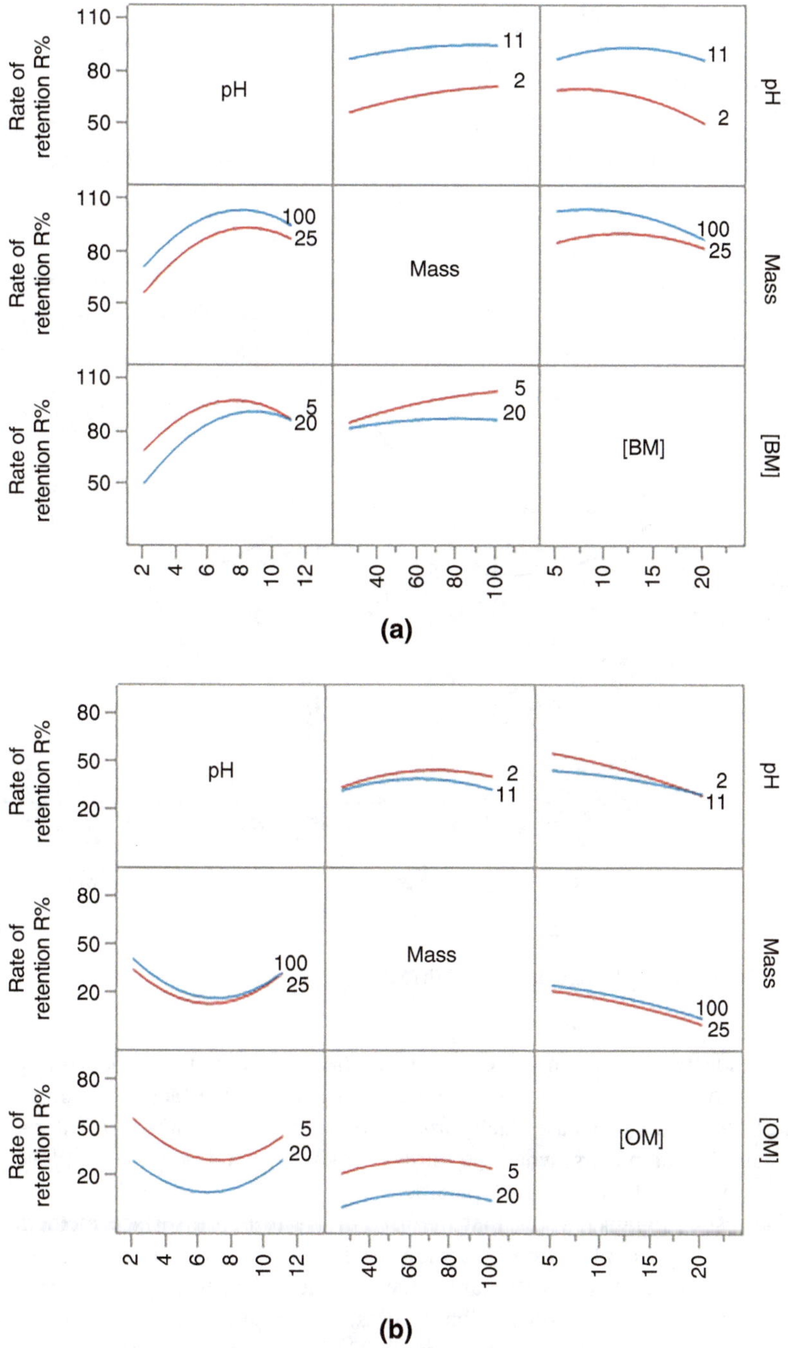

Fig. 4 Interaction profiles of different factors for (**a**) MB and (**b**) MO

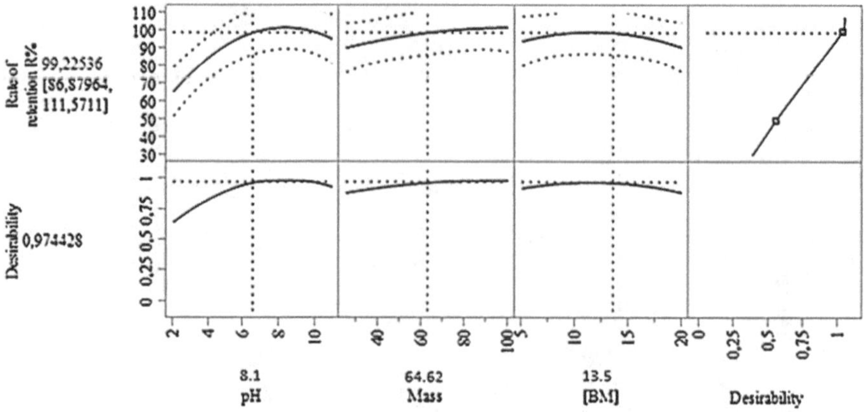

(a)

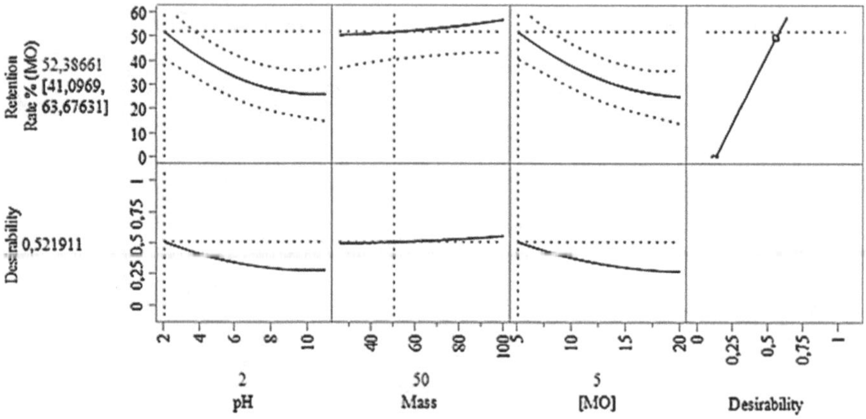

(b)

Fig. 5 Prediction profiler fore (**a**) MB and (**b**) MO

visually locate the zone of the experimental field in which the response is higher and also to determine the various interactions between the factors leading to the desired response. In our case, each curve is a representation of infinity of interactions between two parameters, when the third parameter is fixed in a constant level; the latter is the central level for each parameter.

The results obtained show that the uses of the response surface methodology (RSM) and the central composite design (CCD) make it possible to determine and optimize in a correct and exact manner the influence of the three parameters chosen on the efficiency of the biosorption process. The optimum values of these three parameters give a maximum efficiency of 99.22% and 52.38%. The reliability of the second-order model is based on multiple regressions and tested by the method of analysis (ANOVA). The results show that the models chosen are highly significant

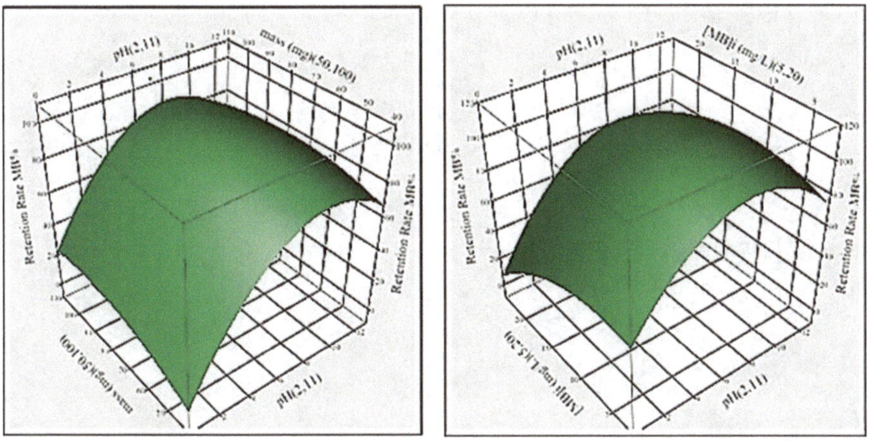

Fig. 6 Three-dimensional response surface for MB

and adequate with the experimental results. The efficiency and performance of adsorption have a negative relationship with the concentration of methyl orange and methylene blue (Figs. 6 and 7), which is in agreement with the literature [29]. This influence is widely studied in the method of adsorption [30]. It has been observed that an increase in the pollutant concentration leads to saturation of the active adsorbent sites and consequently to a decrease in decolonization rates [31]. Unlike the mass of the adsorbent, where the removal efficiency has a positive relationship, this can be explained by the increase in active sites of the material [32]. Finally, the basic pH range between 8 and 10.36 privileges the adsorption of methylene blue; this can be explained by the release of hydroxide ions, which are a good source of hydroxyl radicals HO• [33]. On the other hand, the acidic range of the pH between 4 and 2 privileges the adsorption of methyl orange; this can be explained by the phenomenon of electrostatic attraction, that is to say at acidic pH, the number of sites positively charged increases, which promotes adsorption of anions. The most important effect is that of the adsorbate concentration, followed by the pH, and finally by the adsorbent mass.

17 Conclusions

This work concerns on one hand the valuation of natural waste in the treatment of wastewater by adsorption method and on the other hand the modeling and optimization of the adsorption of MB and MO onto raw date seeds, by the study of the influence of the pH of the aqueous medium, the concentration of OM and MB, and the mass of biosorbent.

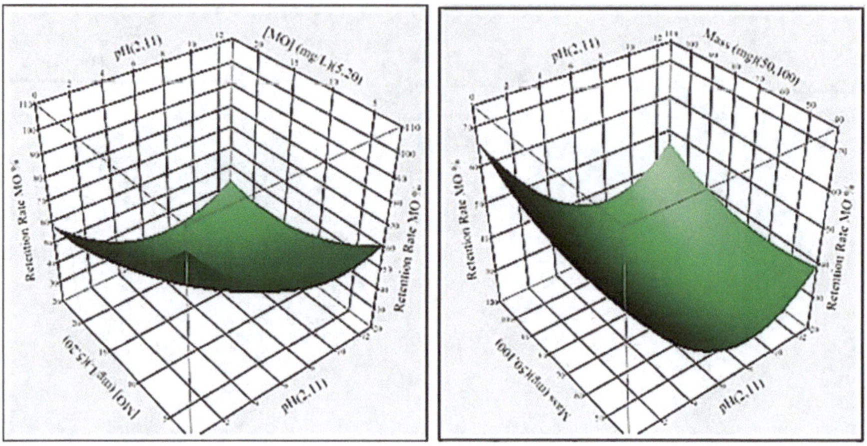

Fig. 7 Three-dimensional response surface for MO

The optimization and modeling of the adsorption of the azo dye MO and MB were performed using an experimental design and more specifically the response surface methodology with a central composite design "CCD."

The regression of the second-order model was tested by the analytical method (ANOVA). This analysis showed that the model is significantly higher and adequate with the experimental results.

The results indicate that all the factors studied influence the adsorption process, and the adsorbent mass is the least pronounced factor in inverse of the effect of the catalyst concentration.

Prediction profiler and storyteller plots indicate that MB adsorption is optimal in basic pH and the optimized parameter values in MB decolonization are pH = 8.1, MB concentration = 13.5 ppm, and mass = 64.62 mg and MO is optimal in acid pH and the parameter values optimized in decolonization of MO are: pH = 2, [MO] = 5 ppm, and mass = 50 mg. These optimal values of the parameters lead to a maximum efficiency of 99.22% and 52.38%, respectively, for MB and MO.

The results could show that the raw date seed used is more effective in removing the dyes.

References

1. Amela, K., Hassen, M. A. and Kerroum, D., Isotherm and Kinetics Study of Biosorption of Cationic Dye onto Banana Peel, Energy Procedia, vol. 19, pp. 286–95, January 1, 2012. DOI: https://doi.org/10.1016/j.egypro.2012.05.208
2. Amini, M., Younesi, H., Bahramifar, N., Lorestani, A. A. Z., Ghorbani, F., Daneshi, A. and Sharifzadeh, M., Application of Response Surface Methodology for Optimization of Lead Biosorption in an Aqueous Solution by Aspergillus Niger, Journal of Hazardous Materials, vol.

154, no. 1, pp. 694–702, June 15, 2008. DOI: https://doi.org/10.1016/j.jhazmat.2007.10.114

3. Arulkumar, M., Sathishkumar, P. and Palvannan, T., Optimization of Orange G Dye Adsorption by Activated Carbon of Thespesia Populnea Pods Using Response Surface Methodology, Journal of Hazardous Materials, vol. 186, no. 1, pp. 827–34, February 15, 2011. DOI: https://doi.org/10.1016/j.jhazmat.2010.11.067
4. Azoulay, K., Bencheikh, I., Moufti, A., Dahchour, A., Mabrouki, J. and El Hajjaji, S., Comparative Study between Static and Dynamic Adsorption Efficiency of Dyes by the Mixture of Palm Waste Using the Central Composite Design, Chemical Data Collections, vol. 27, p. 100385, June 1, 2020. DOI: https://doi.org/10.1016/j.cdc.2020.100385
5. Balcha, A., Yadav, O. P. and Dey, T., Photocatalytic Degradation of Methylene Blue Dye by Zinc Oxide Nanoparticles Obtained from Precipitation and Sol-Gel Methods, Environmental Science and Pollution Research, vol. 23, no. 24, pp. 25485–93, December 1, 2016. DOI: https://doi.org/10.1007/s11356-016-7750-6
6. Bencheikh, I., Azoulay, K., Mabrouki, J., El Hajjaji, S., Dahchour, A., Moufti, A. and Dhiba, D., The Adsorptive Removal of MB Using Chemically Treated Artichoke Leaves: Parametric, Kinetic, Isotherm and Thermodynamic Study, Scientific African, vol. 9, p. e00509, September 1, 2020. DOI: https://doi.org/10.1016/j.sciaf.2020.e00509
7. Bencheikh, Imane, Azoulay, K., Mabrouki, J., El Hajjaji, S., Moufti, A. and Labjar, N., The Use and the Performance of Chemically Treated Artichoke Leaves for Textile Industrial Effluents Treatment, Chemical Data Collections, vol. 31, p. 100597, February 1, 2021. DOI: https://doi.org/10.1016/j.cdc.2020.100597
8. Dean, A., Voss, D. and Draguljić, D., Response Surface Methodology, in Design and Analysis of Experiments, A. Dean D. Voss and D. Draguljić, Eds., Cham: Springer International Publishing, accessed March 31, 2021, from doi:https://doi.org/10.1007/978-3-319-52250-0_16, pp. 565–614, 2017.
9. Dural, M. U., Cavas, L., Papageorgiou, S. K. and Katsaros, F. K., Methylene Blue Adsorption on Activated Carbon Prepared from Posidonia Oceanica (L.) Dead Leaves: Kinetics and Equilibrium Studies, Chemical Engineering Journal, vol. 168, no. 1, pp. 77–85, March 15, 2011. DOI: https://doi.org/10.1016/j.cej.2010.12.038
10. Elmoubarki, R., Taoufik, M., Moufti, A., Tounsadi, H., Mahjoubi, F. Z., Bouabi, Y., Qourzal, S., Abdennouri, M. and Barka, N., Box-Behnken Experimental Design for the Optimization of Methylene Blue Adsorption onto Aleppo Pine Cones, p. 8, 2017.
11. Fakir, L. E., Flayou, M., Dahchour, A., Sebbahi, S., Kifani-Sahban, F. and Hajjaji, S. E., Adsorptive Removal of Copper (II) from Aqueous Solutions on Phosphates: Equilibrium, Kinetics, and Thermodynamics, Desalination and Water Treatment, vol. 57, no. 36, pp. 17118–27, August 1, 2016. DOI: https://doi.org/10.1080/19443994.2015.1112840
12. Fernandes, J. V., Rodrigues, A. M., Menezes, R. R. and Neves, G. de A., Adsorption of Anionic Dye on the Acid-Functionalized Bentonite, Materials, vol. 13, no. 16, p. 3600, January 2020. DOI: https://doi.org/10.3390/ma13163600
13. Genç, N., Durna, E. and Gengec, E., Heterogeneous Activation of Persulfate by Graphene Oxide-TiO2 Nanosheet for Oxidation of Diclofenac: Optimization by Central Composite Design, Water, Air, & Soil Pollution, vol. 229, no. 10, p. 330, September 28, 2018. DOI: https://doi.org/10.1007/s11270-018-3986-8
14. Ghorbani, F., Younesi, H., Ghasempouri, S. M., Zinatizadeh, A. A., Amini, M. and Daneshi, A., Application of Response Surface Methodology for Optimization of Cadmium Biosorption in an Aqueous Solution by Saccharomyces Cerevisiae, Chemical Engineering Journal, vol. 145, no. 2, pp. 267–75, December 15, 2008. DOI: https://doi.org/10.1016/j.cej.2008.04.028
15. Ghoreishi, S. M., Bataghva, E. and Dadkhah, A. A., Response Surface Optimization of Essential Oil and Diosgenin Extraction from Tribulus Terrestris via Supercritical Fluid Technology, Chemical Engineering & Technology, vol. 35, no. 1, pp. 133–41, 2012. doi:https://doi.org/10.1002/ceat.201100091
16. Goupy, J. and Creighton, L., Introduction aux plans d'expériences: Toutes les techniques nécessaires à la conduite d'une étude, 2013.

17. Hadri, M., Chaouki, Z., Draoui, K., Nawdali, M., Barhoun, A., Valdés, H., Drouiche, N. and Zaitan, H., Adsorption of a Cationic Dye from Aqueous Solution Using Low-Cost Moroccan Diatomite: Adsorption Equilibrium, Kinetic and Thermodynamic Studies., Desalination and Water Treatment, vol. 75, pp. 213–24, 2017.
18. Hebeish, A., Ramadan, M. A., Abdel-Halim, E. and Abo-Okeil, A., An Effective Adsorbent Based on Sawdust for Removal of Direct Dye from Aqueous Solutions, Clean Technologies and Environmental Policy, vol. 13, no. 5, pp. 713–18, October 1, 2011. DOI: https://doi.org/10.1007/s10098-010-0343-z
19. Javanbakht, V. and Ghoreishi, S. M., Application of Response Surface Methodology for Optimization of Lead Removal from an Aqueous Solution by a Novel Superparamagnetic Nanocomposite, Adsorption Science & Technology, vol. 35, no. 1–2, pp. 241–60, March 1, 2017. DOI: https://doi.org/10.1177/0263617416674474
20. Khelifi, O., Mehrez, I., Younsi, M., Nacef, M. and Affoune, A. M., METHYL ORANGE ADSORPTION ON BIOSORBENT DERIVED FROM MANGO SEED KERNELS, LARHYSS Journal P-ISSN 1112-3680/E-ISSN 2521-9782, vol. 0, no. 36, pp. 145–56, December 31, 2018.
21. Lemlikchi, W., Drouiche, N., Belaicha, N., Oubagha, N., Baaziz, B. and Mecherri, M. O., Kinetic Study of the Adsorption of Textile Dyes on Synthetic Hydroxyapatite in Aqueous Solution, Journal of Industrial and Engineering Chemistry, vol. 32, pp. 233–37, December 25, 2015. DOI: https://doi.org/10.1016/j.jiec.2015.08.023
22. Liu, Y., Wang, J., Zheng, Y. and Wang, A., Adsorption of Methylene Blue by Kapok Fiber Treated by Sodium Chlorite Optimized with Response Surface Methodology, Chemical Engineering Journal, vol. 184, pp. 248–55, March 1, 2012. DOI: https://doi.org/10.1016/j.cej.2012.01.049
23. Mafra, M. R., Igarashi-Mafra, L., Zuim, D. R., Vasques, É. C. and Ferreira, M. A., Adsorption of Remazol Brilliant Blue on an Orange Peel Adsorbent, Brazilian Journal of Chemical Engineering, vol. 30, no. 3, pp. 657–65, September 2013. DOI: https://doi.org/10.1590/S0104-66322013000300022
24. McKay, G., Otterburn, M. S. and Aga, J. A., Fuller's Earth and Fired Clay as Adsorbents for Dyestuffs, Water, Air, and Soil Pollution, vol. 24, no. 3, pp. 307–22, March 1, 1985. DOI: https://doi.org/10.1007/BF00161790
25. Myers, R. H. and Montgomery, D. C., Response Surface Methodology: Process and Product Optimization Using Designed Experiments, Wiley, pp. 734, 1995.
26. Preparation of Activated Carbon from Phragmites Australis: Equilibrium Behaviour Study, Research Journal of Recent Sciences, vol. 1, no. 8, pp. 10–16, 2012.
27. Radaei, E., Alavi Moghaddam, M. R. and Arami, M., Removal of Reactive Blue 19 from Aqueous Solution by Pomegranate Residual-Based Activated Carbon: Optimization by Response Surface Methodology, Journal of Environmental Health Science and Engineering, vol. 12, p. 65, March 28, 2014. DOI: https://doi.org/10.1186/2052-336X-12-65
28. Roosta, M., Ghaedi, M., Daneshfar, A., Darafarin, S., Sahraei, R. and Purkait, M. K., Simultaneous Ultrasound-Assisted Removal of Sunset Yellow and Erythrosine by ZnS:Ni Nanoparticles Loaded on Activated Carbon: Optimization by Central Composite Design, Ultrasonics Sonochemistry, vol. 21, no. 4, pp. 1441–50, July 1, 2014. DOI: https://doi.org/10.1016/j.ultsonch.2014.01.018
29. Subbaiah, M. V. and Kim, D.-S., Adsorption of Methyl Orange from Aqueous Solution by Aminated Pumpkin Seed Powder: Kinetics, Isotherms, and Thermodynamic Studies, Ecotoxicology and Environmental Safety, vol. 128, pp. 109–17, June 1, 2016. DOI: https://doi.org/10.1016/j.ecoenv.2016.02.016
30. Wu, F.-C., Tseng, R.-L. and Juang, R.-S., Adsorption of Dyes and Phenols from Water on the Activated Carbons Prepared from Corncob Wastes, Environmental Technology, vol. 22, no. 2, pp. 205–13, February 1, 2001. DOI: https://doi.org/10.1080/09593332208618296

31. Yi, F.-Y., Zhu, W., Dang, S., Li, J.-P., Wu, D., Li, Y. and Sun, Z.-M., Polyoxometalates-Based Heterometallic Organic–Inorganic Hybrid Materials for Rapid Adsorption and Selective Separation of Methylene Blue from Aqueous Solutions, Chemical Communications, vol. 51, no. 16, pp. 3336–39, February 10, 2015. DOI: https://doi.org/10.1039/C4CC09569C
32. Zhang, H., Shi, X., Li, J., Kumar, P. and Liu, B., Selective Dye Adsorption by Zeolitic Imidazolate Framework-8 Loaded UiO-66-NH2, Nanomaterials, vol. 9, no. 9, accessed March 31, 2021, from https://www.ncbi.nlm.nih.gov/pmc/articles/PMC6781039/, September 8, 2019. DOI: https://doi.org/10.3390/nano9091283
33. Azoulay, K. Bencheikh, I. 1 Mabrouki, J. Samghouli, N. Moufti, A. Dahchour, A. and El Hajjaji, S., Adsorption mechanisms of azo dyes binary mixture onto different raw palm wastes, International Journal of Environmental Analytical Chemistry, 1–20, 2021. https://doi.org/10.1080/03067319.2021.1878165

Mathematical and Statistical Study for the Wastewater Adsorbent Regeneration Using the Central Composite Design

Imane Bencheikh, Karima Azoulay, Nora Samghouli, Jamal Mabrouki, Loubna Bouhachlaf, Ahmed Moufti, and Souad El Hajjaji

1 Introduction

The experimental designs are based on organizing experiments in order to obtain reliable results economically and efficiently. In recent years, they have become very applied in optimizing experimental environmental research, such as wastewater treatment. Because they allow a simple interpretation of the results and a prediction of the behavior of the studied system for unexperienced values of the parameters by using the statistical tool 'Standard ISO 3534-3' [1, 2].

Centered composite designs are widely used in this kind of study. They have three parts:

- The two levels (−1, +1) of the complete factorial design, 2*f* experiments (*f*, number of factors)
- The two levels ($-\alpha$, $+\alpha$) of the star plane, 2*f* experiments
- The point in the center of the study area

I. Bencheikh (✉) · K. Azoulay · N. Samghouli · J. Mabrouki · L. Bouhachlaf · S. El Hajjaji
Laboratory of Spectroscopy, Molecular, Modelling, Materials, Nanomaterials, Water and Environment, (LS3MN2E), Department of Chemistry, Faculty of Sciences, Mohammed V University in Rabat, Rabat, Morocco
e-mail: imane-bencheikh@um5r.ac.ma

A. Moufti
Equipe Science de l'Environnement et Matériaux Appliqué (SEMA), Université Sultan Moulay Sliman, Khouribga, Morocco

Environmental Sciences and Applied Materials (ESAM) Team, khouribga ply-disciplinary Faculty (FPK), Khouribga, Morocco

Regional Center for Careers in Education and Training, Casablanca-Settat, Morocco

M. Azrour et al. (eds.), *IoT and Smart Devices for Sustainable Environment*, EAI/Springer Innovations in Communication and Computing,
https://doi.org/10.1007/978-3-030-90083-0_6

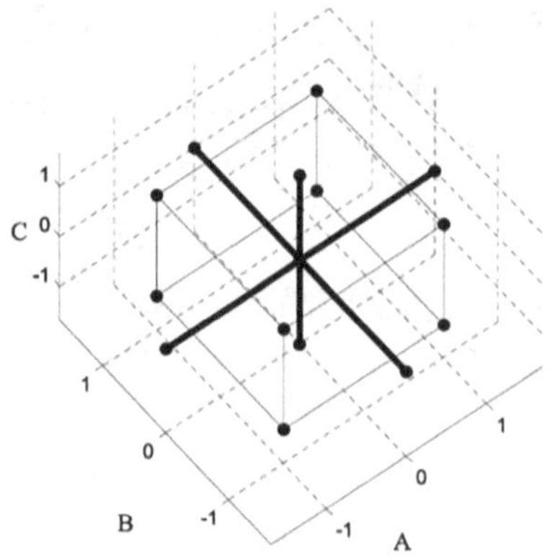

Fig. 1 Geometric location of the experimental points of a composite plane centered with three factors

The total number of N tests to be performed for f factors is given by Eq. 1:

$$N = 2f + 2f + 1 \quad (1)$$

The total number of N tests to be performed for f factors is given by Eq. 2:

$$N = 2f + 2f + 1 \quad (2)$$

The value of α must be equal to: α = (2*f*) 1/4 (Eq. 1). These designs have the advantage of allowing a sequential conduct of a study and requiring a relatively small number of experiments [3]. The composite design for three factors has 15 experiments. The geometrical locus of the experimental points is represented in Fig. 1 [2, 4].

The exploitation of the results obtained by the software is often quite fast. Among the parameters to be analyzed, we have determination coefficients (R^2 and R^2 adjusted), which measure the quality of the model. It takes values between 0 and 1. If the value of R^2 is close to 1, it indicates that the model is able to represent the values of the responses measured experimentally. If it is equal to 0, it means that the model is not adequate for the explanation of the studied phenomenon [2, 4]. We also have the analysis of variance, which is an essential tool in the determination of the significant effects or a mathematical model and whose principle is based on the calculation of the total difference between the different measurements *yi* of the experiments design and the average of these measurements. This deviation can be subdivided into two: regression deviation, which represents the part that can be modeled by the mathematical model, and residual deviation, which constitutes the part that cannot be modeled by the model. This last difference constitutes the difference between the response measured experimentally and that obtained by modeling. It is therefore the sum of the adjustment deviation and the pure experimental deviation [2, 4]. In addition, the probability P is a value that occurs in the analysis of variance, whose value is between 0 and 1. It mainly gives an idea about the significance of the studied factors. If the *P*-value of a factor *i* is less than 0.05, it means that the factor is significant. If the *P*-value of a factor *i* is greater than 0.05, it means that the factor is not significant; if it is less than 0.01, we can conclude that the factor is highly significant [2, 4]. Finally, we have the prediction profiler, which allows us to study and visualize the impact of changes for the factors in function of the predicted values and on the degree of our desirability. The lines of the profile illustrate the power and the evolution of the impact of the modifications of the factors on the values of the studied response. Each modification made on the profiler makes the line of the factors studied more or less inclined, allowing instant visualization of the impact of the parameters on the studied response [2, 4].

2 Method of Computation

The process of regeneration and reuse of the adsorbent is essential for the adsorption to be economically viable [5]. In this study, the choice of regenerating solvents was made based on a literature research taking into account the mixture of pollutants (organic and inorganic) eliminated by our support (FA-T) [6, 7]. The eluents chosen for this purpose are ethanol-water, HCl, NaOH, NaCl, and distilled water. After adsorption, the adsorption column, containing the adsorbent–adsorbate was washed with distilled water to remove all traces of the residual adsorbate (s) on the surface material and then washed with the mentioned solvents separately. The various parameters, which affect the regeneration, were studied using CCD, and the experimental conditions for regeneration are described in Tables 1 and 2. Finally, the column is rewashed with distilled water. The regenerated adsorbent is then used for readsorption tests. The efficiency of regeneration in terms of removal of a pollutant *i* or in a system *y* (ER_i) is calculated by Eqs. 4 and 5 (Table 3).

Table 1 Experimental domain of FA-T regeneration by eluents (ethanol-water, HCl, NaOH, and NaCl)

Variables	Symbols	Level	
		−1	+1
[Eluent] (M)	X_1	0.05	0.5
Regeneration time (min)	X_2	30	120
Flow (mL/min)	X_3	1	10

Table 2 Experimental domains of FA-T regeneration by distilled water

Variables	Symbols	Level	
		−1	+1
Regeneration time (min)	X_1	30	120
Flow (mL/min)	X_2	1	10

Table 3 The equations of regeneration efficiency in terms of reelimination of a pollutant *i*: studied pollutants: OM. Cu(II)

Equations
$ER_i\ (\%) = \frac{\text{Adsorption capacity oth the regenerated adsorbent}}{\text{Adsortpion capacity of the fresh adsorbent}} * 100$ (4)
$ER_{Conductivity}\ (\%) = \frac{\text{The percentage of conductivity reduction using theregenerated adsorbent}}{\text{The percentage of conductivity reduction using the fresh adsorbent}} * 100$ (5)

ER_{OM}: The adsorbent efficiency in terms of removing organic matter (OM) after regeneration (%)
$ER_{cu(II)}$: The adsorbent efficiency in terms of removing copper in solution (Cu (II)) after regeneration (%)
$ER_{Conductivity}$: Percentage of the conductivity reduction after the adsorbent regeneration (%)

3 Results and Discussion

3.1 Model Validation

Figure 2 gathers the retention rate values predicted by the centered composite plane and those obtained experimentally for the three responses studied (COD (OM), Cu (II), and conductivity).

According to Fig. 2, the estimated values of the determination coefficient for ethanol-water are R^2 Ethanole-water = 0.99, R^2 Ethanole-water = 0.98, and R^2 Ethanole-water = 0.97, respectively, for COD, Cu (II), and conductivity; the estimated values of the determination coefficient for HCl are R^2 HCl = 0.97, R^2 HCl = 1, and R^2 HCl = 0.98, respectively, for COD, Cu (II), and conductivity; the estimated determination coefficient values for NaOH are R^2 NaOH = 0.99, R^2 NaOH = 0.99, and R^2 NaOH = 1, respectively, for COD, Cu (II), and cosnductivity; the estimated determination coefficient values for NaCl are R^2 NaCl = 0.96, R^2 NaCl = 0.99, and R^2 NaCl = 0.98, respectively, for COD, Cu (II), and conductivity; and the estimated determination coefficient for distilled water are R^2 distilled water = 0.99, R^2 distilled water = 0.89, and R^2 distilled water = 1 for COD, Cu (II), and conductivity, respectively. These values indicate that there is a good accordance between the experimental and predicted values of the FA-T regeneration by ethanol-water, HCl, NaOH, NaCl, and distilled water in terms of retreatment of OM, Cu (II), and conductivity. This shows that the model is relatively well adjusted. Tables 4 and 5 group the values of the different studied factors' terms for the different adsorbent–adsorbate systems.

Knowing that the most influential factors on the regeneration efficiency are those whose terms have a *p*-value <0.05. Therefore, from Table 5, we note that the concentration of ethanol-water, the total time of regeneration, the interaction between the ethanol-water concentration and the total regeneration time, and the double interaction of ethanol-water have more effect on the regeneration of FA-T by ethanol-water in terms of adsorption of OM, Cu (II). On the other hand, the most influencing parameters on the regeneration of FA-T by ethanol-water in terms of retreatment of conductivity are the ethanol-water concentration, the total time of regeneration, and the double interaction of the ethanol-water. Also, the [HCl], total

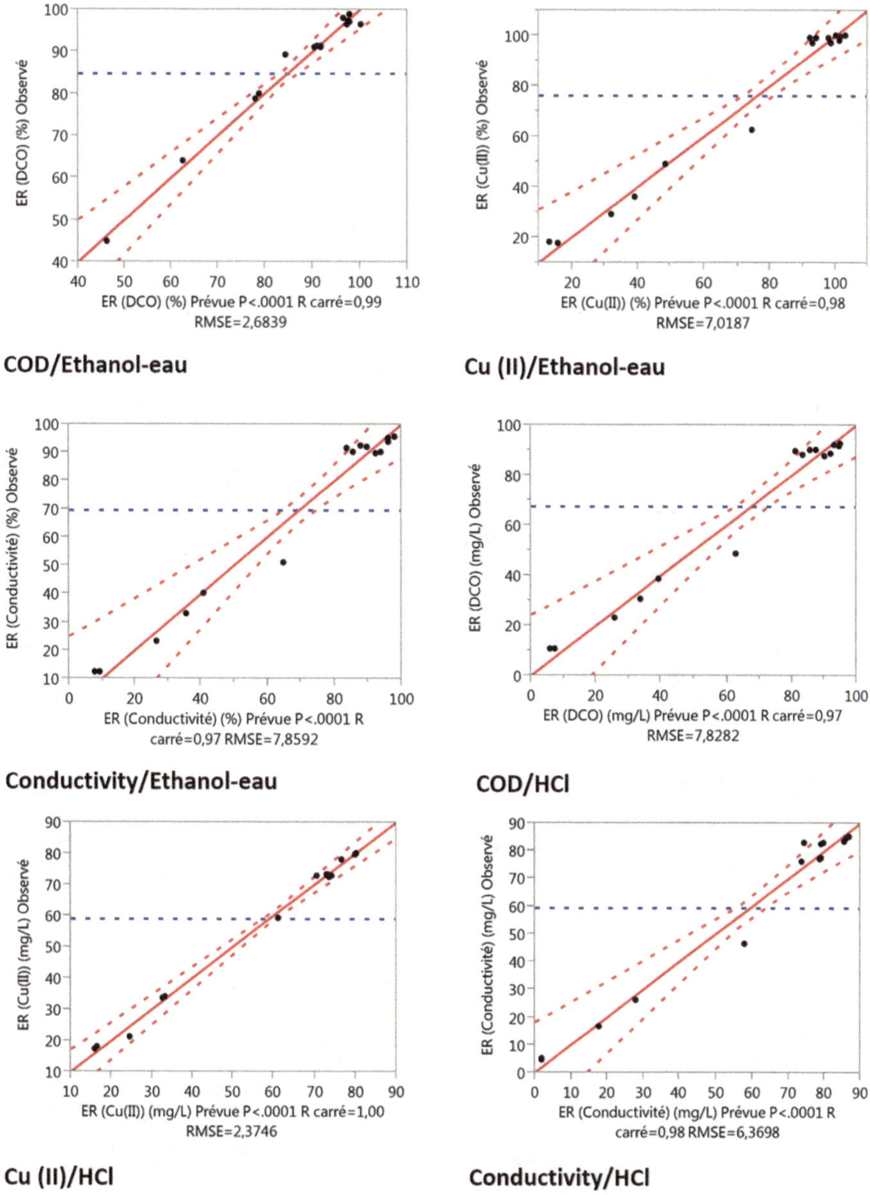

Fig. 2 Graphs of the values observed versus the predicted values for the regeneration of FA-T by ethanol-water, HCl, NaOH, NaCl, and distilled water in terms of readsorption of OM, Cu reduction of conductivity

regeneration time, and the interaction between [HCl] and flow rate have a significant effect on the efficiency of FA-T regeneration in terms of conductivity reduction.

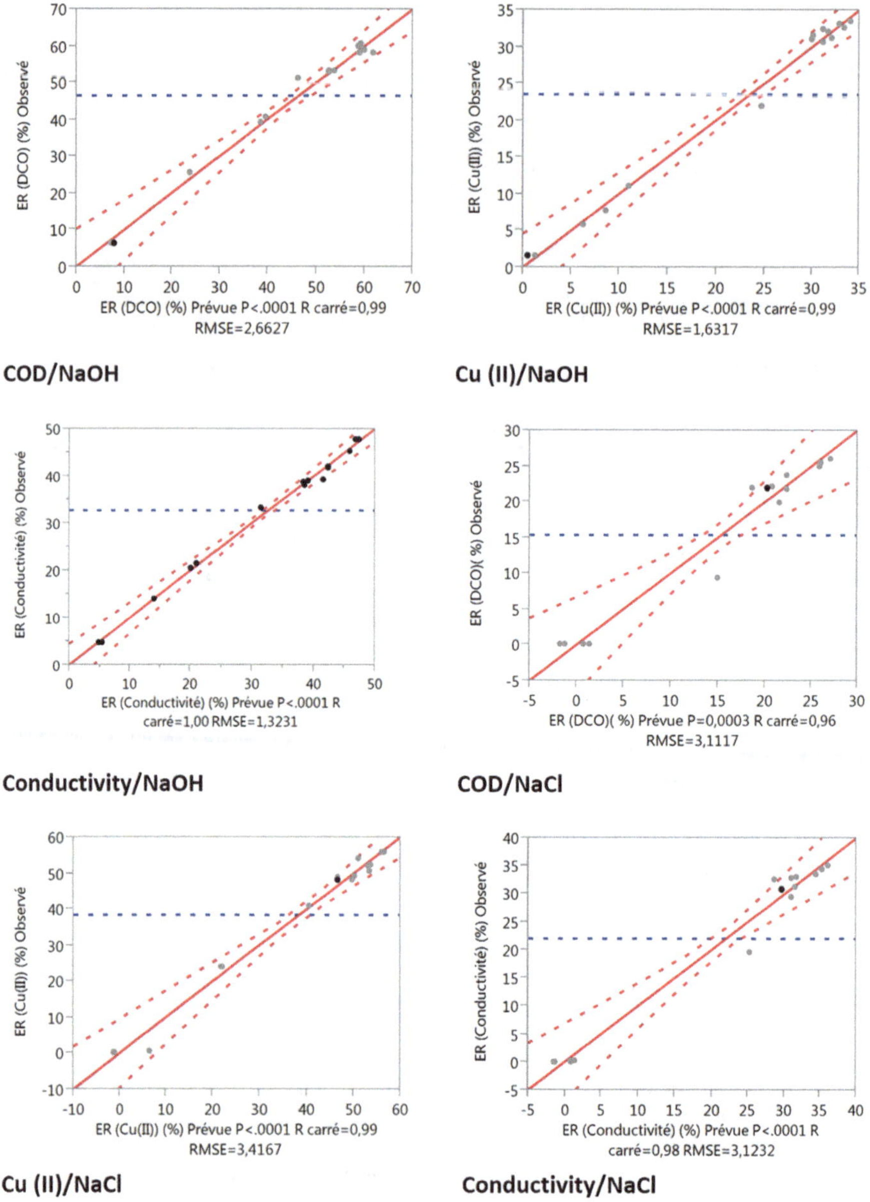

Fig. 2 (continued)

However, we also have the effect of the double interaction of total regeneration time for the regeneration of FA-T by HCl in terms of COD retreatment and the effect of the interaction between [HCl] and the total regeneration time for the regeneration of FA-T by HCl in terms of Cu (II) adsorption. Concerning the regeneration by

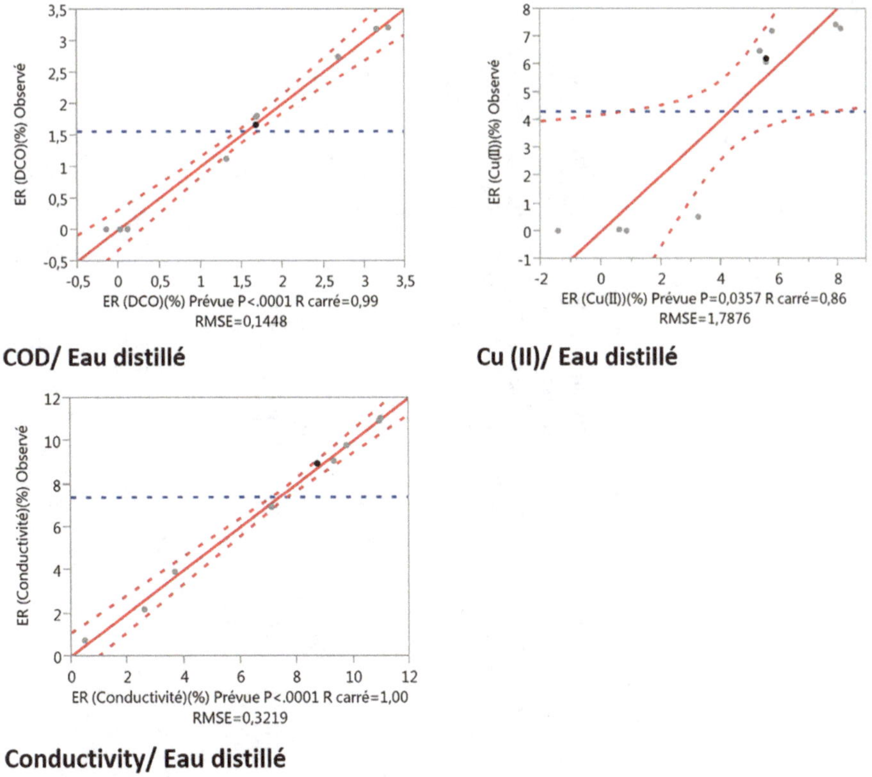

Fig. 2 (continued)

NaOH, we have four terms, which are statistically significant in the regeneration of FA-T by NaOH in terms of retreatment of OM, Cu (II), and conductivity and which are: [NaOH], the total time of regeneration, the interaction between [NaOH], the total regeneration time, and the double interaction [NaOH]. We also have the NaCl concentration and the double interaction of the NaCl concentration for the regeneration of FA-T by NaCl in terms of OM retreatment and conductivity. The NaCl concentration, total regeneration time, the interaction between [NaCl] and total regeneration time and the double interaction of [NaCl] are the most influencing parameters for the regeneration of FA-T by NaCl in terms of retreatment of Cu (II).

From Table 5 we found that the effects of all the factors and their interaction are significant for the regeneration of FA-T by distilled water in terms of conductivity retreatment, while the effect of total regeneration time and flow rate are significant for the regeneration in terms of COD abatement; on the other hand, only the effect of total regeneration time is significant for the regeneration of FA-T in terms of Cu retreatment (II).

Table 4 FA-T by the different

Term	Constant	X_1	X_2	X_3	X_1*X_2	X_1*X_3	X_2*X_3	X_1*X_1	X_2*X_2		X_3*X_3
COD	Ethanoleau	<0.0001*	<0.0001*	<0.0001*	0.7073	<0.0001*	0.9241	0.8440	0.0002*	0.7087	0.5837
Cu(II)		<0.0001*	<0.0001*	0.0056*	0.5491	0.0479*	0.5381	0.5241	0.0005*	0.0575	0.1540
Conductivity		<0.0001*	<0.0001*	0.0067*	0.7366	0.0991	0.7761	0.7545	0.0016*	0.0555	0.1550
COD	HCl	<0.0001*	<0.0001*	0.0070*	0.7120	0.0861	0.7539	0.7434	0.0018*	0.0452*	0.1667
Cu(II)		<0.0001*	<0.0001*	<0.0001*	0.7470	0.0224*	0.9108	0.9817	<0.0001*	0.0511	0.0591
Conductivity		<0.0001*	<0.0001*	0.0044*	0.9511	0.0746	0.9466	0.9970	0.0007*	0.0898	0.2339
COD	NaOH	<0.0001*	<0.0001*	<0.0001*	0.9936	<0.0001*	0.7079	0.8618	0.0002*	0.8034	0.5603
Cu(II)		<0.0001*	<0.0001*	0.0011*	0.3984	0.0196*	0.6034	0.4990	<0.0001*	0.0369*	0.1522
Conductivity		<0.0001*	<0.0001*	<0.0001*	0.5781	0.0008*	0.8792	0.8308	<0.0001*	0.0672	0.5288
COD	NaCl	<0.0001*	<0.0001*	0.1094	0.6887	0.5511	0.8191	0.8928	0.0099*	0.1116	0.4075
Cu(II)		<0.0001*	<0.0001*	0.0006*	0.8752	0.0042*	0.9817	0.9912	<0.0001*	0.8465	0.1385
Conductivity		<0.0001*	<0.0001*	0.1285	0.7977	0.7230	0.9286	0.9582	0.0004*	0.2108	0.4114

Table 5 *p*-value of terms related to the regeneration efficiency of FA-T by distilled water in terms of retreatment of OM, Cu (II), and conductivity

Term	COD	Cu (II)	Conductivity
Constant	<0.0001*	0.0017*	<0.0001*
X_1 (min)(10,120)	<0.0001*	0.0041*	<0.0001*
X_2 (mL/min)(1,10)	0.0229*	0.2097	0.0004*
X_1 (min)*X_2(mL/min)	0.1730	0.9554	0.0281*
X_1 (min)*X_1 (min)	0.6746	0.3790	0.0002*
X_2 (mL/min)*X_2 (mL/min)	0.1285	0.3100	0.0432*

X_1: Total regeneration time; X_2: Flow; *: the value is below 0.05

4 Prediction Profiler

Figure 3 gathers the profilers for predicting the optimal conditions of the solvent concentration (ethanol-water, HCl, NaOH, and NaCl), the time and flow rate of the solvent purge to have the best possible regeneration of the FA-T, capable of re-adsorb OM molecules and Cu (II) ions, and decrease the conductivity value with high efficiency.

From Fig. 3a it can be seen that the rate of regeneration increases as the concentration of ethanol-water and the total regeneration time increase. In addition, the retention rate is almost stable depending on the variation in flow. The optimal conditions are: 0.3 M for the [ethanol-water], 90 min for total regeneration time, and 3 mL/min of flow rate with desirability close to 1.

According to Fig. 3b, we noticed the same rate of variation is observed for HCl with optimal conditions of the regeneration more or less different: [HCl] = 0.329 M, time = 61.7 min, flow rate = 1 mL/min with a desirability value close to 1 and which corresponds to the maximum yield values up to 96.7%; 78%, 60%, and 85.15%, respectively for COD, Cu (II), and conductivity.

We also noticed from Fig. 3c that there is an increase in the regeneration yield for FA-T in terms of retreatment of OM, Cu (II), and conductivity using NaOH, which is observed with the increase of NaOH concentration of time. On the other hand, increasing the flow rate does not improve the efficiency of the regeneration of FA-T by NaOH in terms of retreatment of OM, Cu (II), and conductivity. The optimum conditions for regeneration are [NaOH] = 0.374 M; total regeneration time of 79.3 min and a flow rate of 1.09 mL/min. The maximum efficiency observed under these conditions is 59.39% for COD abatement, 34.45% for Cu (II), and 44.88% for conductivity.

For NaCl, we find that a maximum removal of 24%, 55%, and 35% for COD, Cu (II), and conductivity, respectively, is obtained by adding 0.356 M of [NaCl], after 60 min with a flow rate of 1 mL/min.

For distilled water, we notice that the regeneration efficiency remains low even with increasing flow and time.

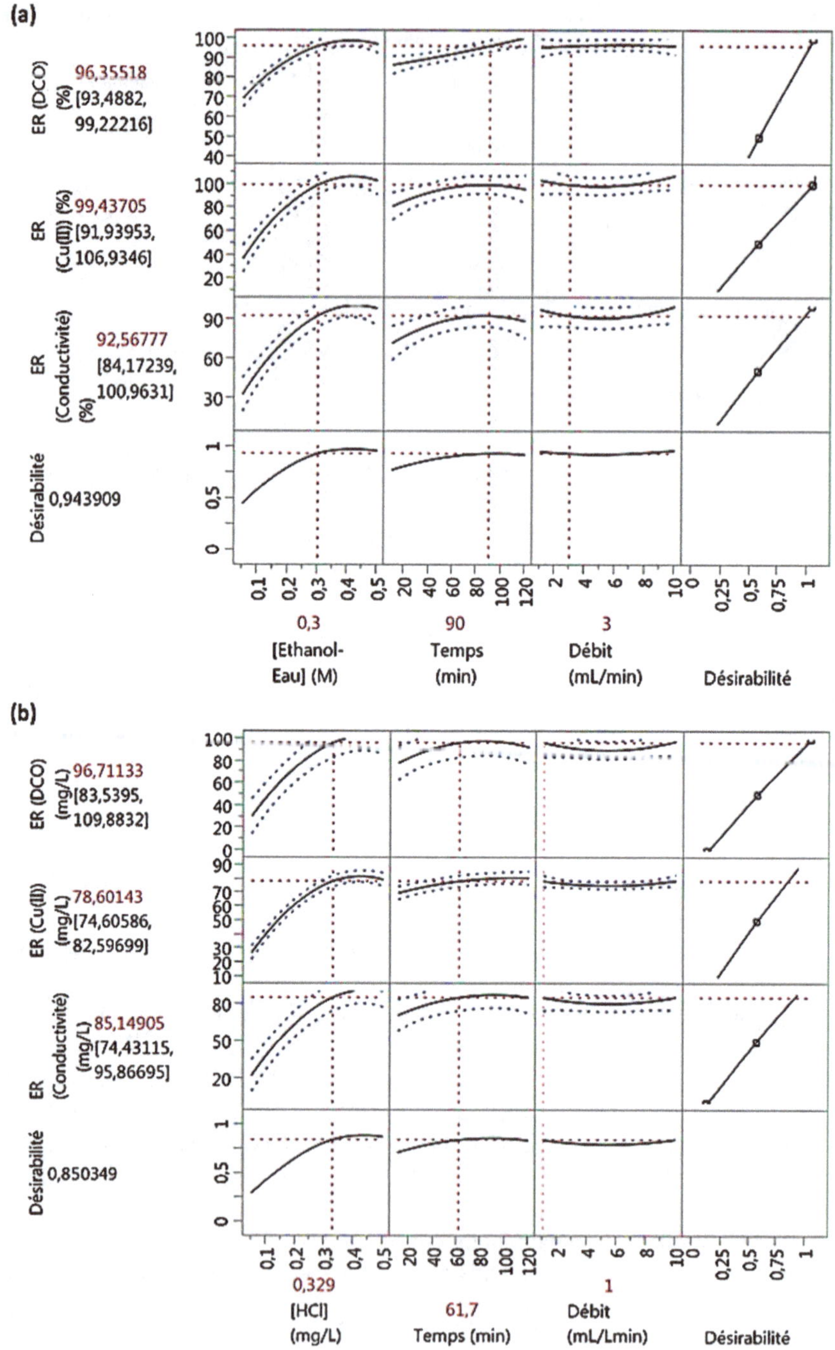

Fig. 3 Prediction profiler of the regeneration of FA-T by ethanol-water, HCl, NaOH, NaCl, and distilled water in the different systems studied

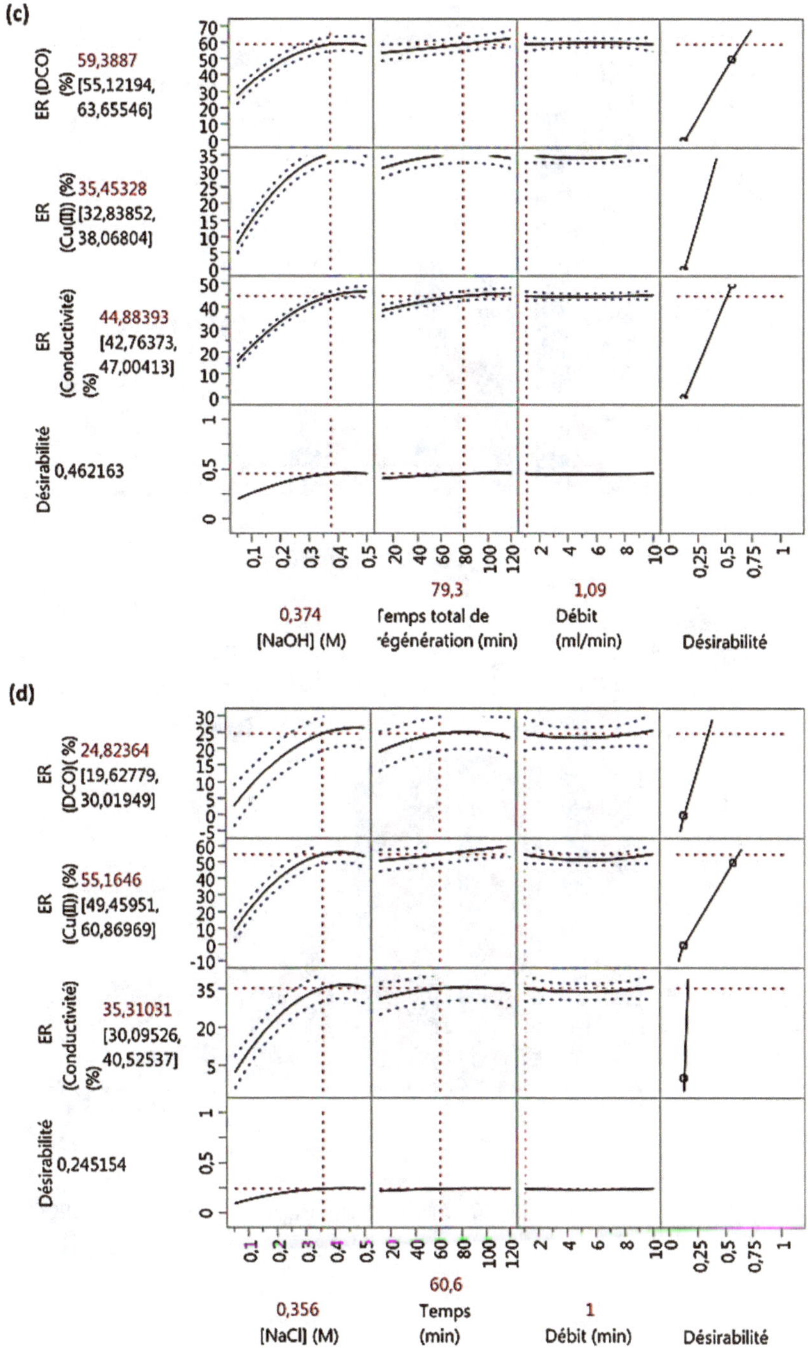

Fig. 3 (continued)

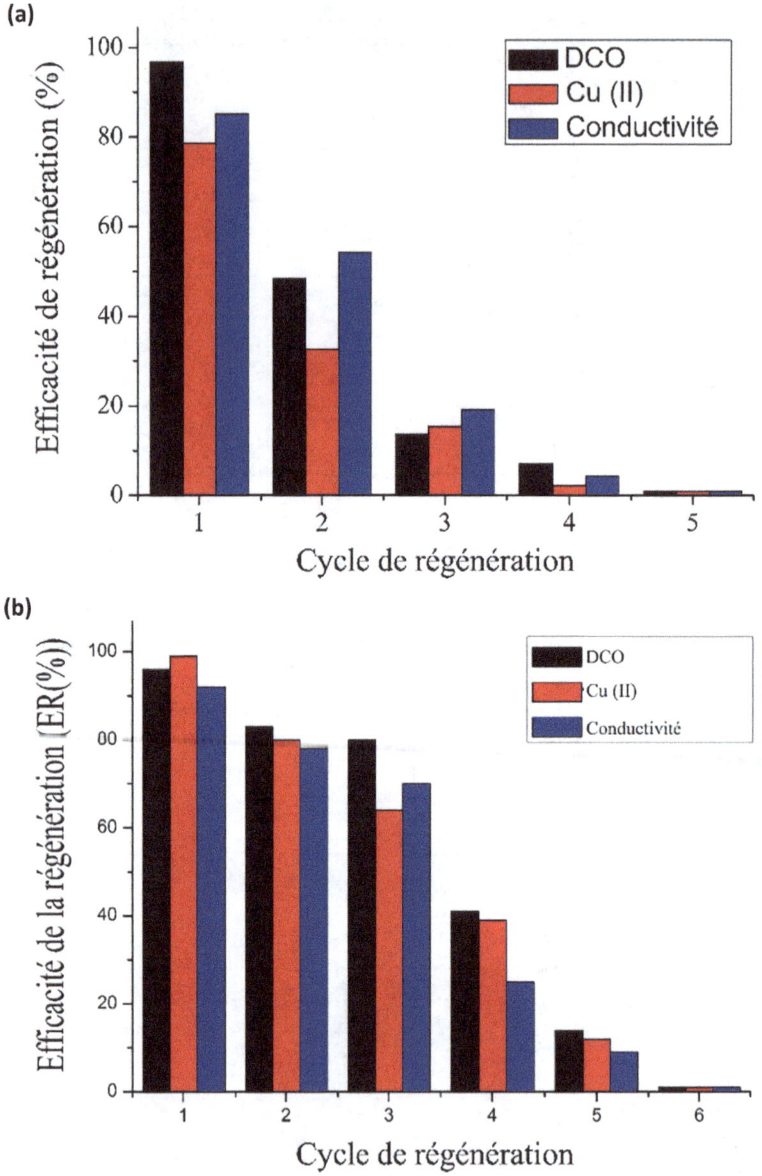

Fig. 4 FA-T regeneration cycles with HCl (**a**) and ethanol-water (**b**) for reprocessing of COD, Cu (II), and conductivity

4.1 Regeneration Cycle

To establish the most efficient eluent, a comparison between the numbers of the regeneration cycles for FA-T by each eluent was carried out (Fig. 4).

From Fig. 4, we notice that the number of efficient regeneration cycles using HCl is less compared with ethanol-water whose regeneration remains efficient until the third cycle with values ranging from 99% to 56% for different pollutants studied.

5 Conclusion

The results of the regeneration study showed that the FA-T can be reused practically up to five cycles with a regeneration efficiency ranging from 99% to 52% using ethanol-water as solvent in the case of synthesized textile solution and up to three cycles for real textile water with efficiency ranging from 99% to 56% for the different pollutants studied (OM and Cu (II)). The difference observed between the number of cycles in the synthetic and real medium is relative to the initial pollutant load of the two mediums. Also the drop in performance observed between cycles is caused by the destruction of active sites during chemical regeneration and which causes a loss of performance of the material.

References

1. K. Sandrine, Application de la méthodologie des plans d'expériences et de l'analyse de données à l'optimisation des processus de dépôt, Thèse de doctorat, Université de limoges, faculté de des Sciences et Techniques, France, (2004) 234.
2. L. El fakir, Adsorption de colorants de textiles et d'un métal lourd (Cu) sur des matériaux naturels et synthétiques: modélisation et études cinétique et thermodynamique, rapport de thèse de doctorat, Université Mohammed V, Faculté des Sciences, Rabat, 2017.
3. E. Chalmin, Caracterisation des oxydes de manganese et usage des pigments noirs au paleolithique superieur, thése de doctorat, Université de Marne-La-Vallée, Institut Francilien des Géosciences, (2006) 379
4. J. Goupy, Tutoriel Les Plans D'experiences, Revue Modulad, (2006) 34.
5. B.J. Dotson, R.L. Slobod, P.N. McCreery, et J.W. Spurlock. Porosity-measurement comparisons by five laboratories. Society of Petroleum Engineers. Vol 73. p. 341–346. (1951). doi:https://doi.org/10.2118/951341-G.
6. B. Sancey, N. Morin-Crini, L. F Lucas, F. Degiorgi, J.F. Minary, P. M. Badot, et G. Crini. La bioadsorption sur amidon réticulé pour enlever des métaux des effluents industriels. Revue des sciences de l'eau. Vol. 23. p. 275–287. 2010. doi:https://doi.org/10.7202/044689ar.
7. L. Minghua, D. Yun, Z. Huaiyu, et Z. Xinshen. Adsorption and desorption of copper (II) from solutions on new spherical cellulose adsorbent, Journal of Applied Polymer Science. Vol. 84. p. 478–485. 2002. doi:https://doi.org/10.1002/app.10114

IoT and Reality Mining for Real-Time Disease Prediction

Hiba Asri

1 Introduction

It is definitely true that big data enclose multiple sectors, including health care, education, economics, agriculture, and industry. As a matter of fact, a big number of information are produced every short time in all disciplines cited before [1, 31]. Therefore, we need performance tools to make this gigantic amount of data valuable knowledge. Machine learning and data mining tools represent perfect approaches to deal with this fact [28, 29].

After all, by combining techniques of big data analytics, machine learning algorithms, and IoT, reality mining has the capability to collect all data related to our behaviors: individual or collective ones. In this context, we propose a new predictive and preventive system of diseases to make people's lives healthier by including prevention and prediction of outcomes earlier [26]. This new system is developed using data mining algorithms and big data predictive analytics. All data used are collected from mobile phones and wearable healthcare sensors.

During our study, we face multiple challenges: (1) big data analytics tools and reality mining applied to smartphones and healthcare sensors' data to predict patterns; (2) propose a new solution, which can be applied to any data prediction case study such as breast cancer, heart diseases, stress, weather, obesity, depression, among others; (3) use real-time data from healthcare sensors and mobile phones; and (4) have accurate results and predictions because of the critical field we work on.

H. Asri (✉)
LIMA Laboratory, Department of Computer Science, National School of Applied Sciences, Ibn Zohr University, Agadir, Morocco
e-mail: h.asri@uiz.ac.ma

M. Azrour et al. (eds.), *IoT and Smart Devices for Sustainable Environment*, EAI/Springer Innovations in Communication and Computing,
https://doi.org/10.1007/978-3-030-90083-0_7

In this chapter, we introduce existing works in the healthcare literature about predicting outcomes/patterns. We then present an overview of the proposed prediction system and the prediction process we conducted. In another point, we illustrate the process of collecting data, actors of the system, the flow of information between all parts, and the integration and the implementation of IoT systems in the present work. In addition, we discuss data conceptual modeling and implementation in terms of big data management tool and conception method. Finally, we conclude this chapter with a case study of real-time miscarriage prediction system that benefits from big data tools, IoT systems, and machine learning algorithms. Real-world data of real pregnant women are collected and used during this experience.

2 Data Mining for Predictive Analytics: Background

In literature, we can find many researches about the use of machine learning and data mining techniques in predicting patterns. Those researches include many fields such as healthcare, agriculture, education, environment, industry, among others [30, 33].

Authors in [2] explained the power of big data technologies in the healthcare sector. Acccording to their study, the use of big data analytics tools, reality mining and biomedical informatics; contributed positively in predicting outcomes and reducing healthcare delivery costing.

Authors in [3] use data mining to classify articles to customer relation management (CRM). The data represent an academic database between 2000 and 2006 and contain 87 papers [3]. Authors distinguish CRM categories by comparing seven data mining techniques. Through this analysis, the authors propose a guide for future researchers applied to CRM.

Authors in [4] describe a new approach based on a personal communication system. The goal of this study is to predict the next intercell of a mobile's activity through three phases: history of user's information, pattern's rules, and outcome prediction. Comparing with two other prediction models, the new algorithm shows its efficiency in terms of precision and recall.

In a study [5], the authors test five data mining algorithms applied to different wind farm datasets collected from 100 turbines. The goal of this work is to predict the power of a wind farm at different time range. As a result, the support vector machine (SVM) performs well in terms of positive predictions of wind speed and wind power at an interval of 10 min up to 1 h into the future, while the multilayer perceptron algorithm carries out over hour-long ranges, up to 4 h, ahead.

To set up a new efficacious predictive and preventive analytics system, we have to benefit from performance tools that can analyze the enormous data collected such as advanced big data analytics algorithms, data mining, IoT, and intelligent systems. In addition to all techniques cited before, our proposed system includes real-world data gathered from healthcare sensors and mobile phones. This particularity will allow the final system's users to be more concerned and to constitute a real benefit for prediction. Thus, the new system architecture grants real-time predictions and proactively proposes recommendations.

3 The Prediction System

The assets of using mobile phones and wearable healthcare sensors are providing both real-world data on real-time and contextual information. In fact, a person's activities can be detected, thanks to the accelerometer sensor included in the smartphone. In addition, we can determine people's location using the GPS tool in the mobile phone. Furthermore, additional data, like age, gender, or weight, can be collected through a registration form of a mobile application that we created. On the other hand, thanks to the use of wearable sensors, healthcare information can be collected. In fact, we were able to collect healthcare data, such as body temperature, heart rate, stress, activity level, sleep activity, and other healthcare data that help in predicting probable miscarriage.

The originality of the proposed system is the use of real-time world contextual measurements coming from the wearable sensors and mobile phones to help forecasting. As a result, actors receive specific information depending on the profile of the user.

The major features of our system can be summarized as follows (see Fig. 1):

- Real-world data collected on real-time from wearables and mobile phones: location, activity, sleep activity, stress, heartbeat, body temperature, among others.
- Pattern recognition based on real-time data and registered data.
- Effective feedback: Actors receive relevant information based on their behavior.
- Real-world data and database size.

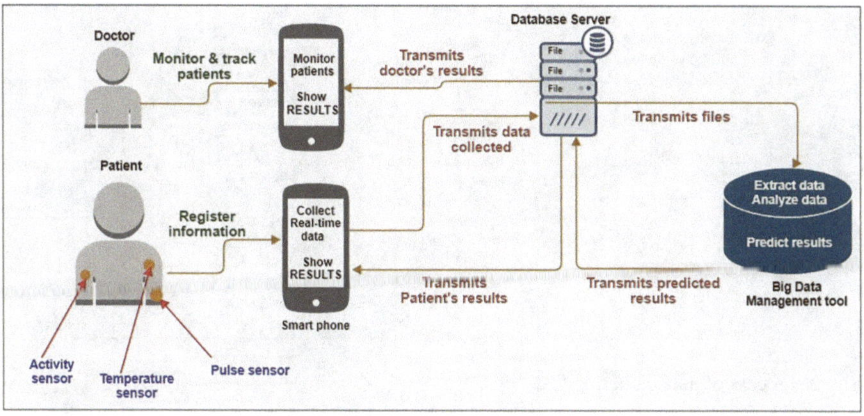

Fig. 1 Prediction system overview

3.1 *Reality Mining for Prediction*

3.1.1 Gathering Process Overview

In prediction situations, the process of gathering data is crucial because outcomes depend on inputs collected. And especially when dealing with life or death of the person, facts must be accurate and error-free. In our study, we gather data in three different ways:

- Data from users' profiles. These data can be updated during the process.
- Real-time data from mobile phones, such as activity, location, voice call, and text.
- Real-time data from wearables.

Figure 2 expresses the data gathering process.

3.1.2 IoT Implementation

Sensors and Tools

Sensors represent the source of IoT data and become up to date. Nowadays, we can find out data that were previously impossible. In this work, multiple IoT health sensors were used to measure features, namely, temperature variation (TV), blood pressure (BP), heart rate variability (HRV), emotion, and activity (see Table 1).

Many robotic tools exist for supervising and processing data coming from sensors, we decide to work with:

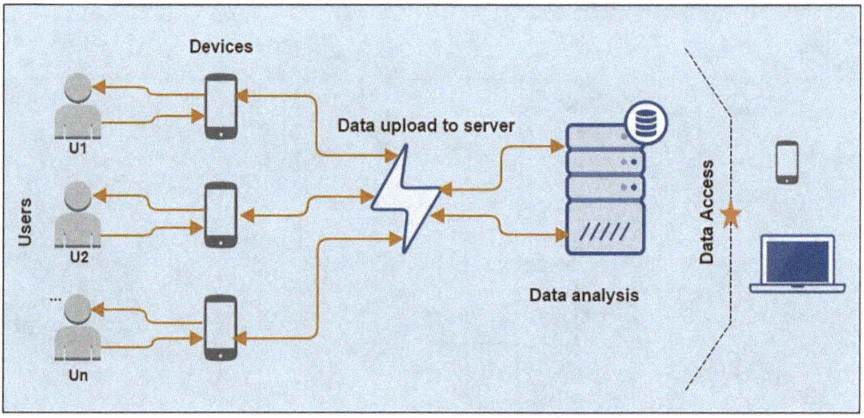

Fig. 2 Process of data gathering

Table 1 Gathering sensors data

Sensor	Measurements
Pulse sensor[a]	Heart rate variability, emotion and stress/blood pressure
Temperature sensor[b]	Body temperature variation
Acceleration sensor[c]	Activity level (high, medium, low)

[a]Heart Rate Pulse sensor for Arduino/Raspberry Pi
[b]I2 C infrared temperature sensor mlx90614 for Arduino Raspberry Pi/ 0.1 °C Accuracy
[c]Kwmobile acceleration sensor with digital output, 3-axis gyroscope for Arduino, Genuino and Raspberry Pi

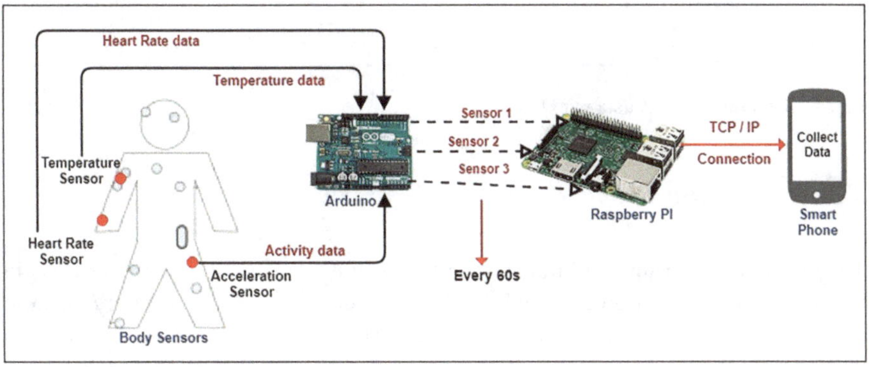

Fig. 3 Gathering sensors data workflow

- Arduino (an open-source platform) for gathering data from health sensors.
- Raspberry Pi (a single-board computer) for programming and processing.

System Implementation

According to the architecture of our proposed system, all patients are equipped with all sensors described before. Arduino is used to collect data every 3 s. Data are stored in JSON file and then to be remitted to Raspberry Pi for processing (see Fig. 3). Once the process is finished, we analyze the data using machine-learning algorithms and big data platforms. This process is repeated every 60 s.

In the other parts of the IoT architecture, mobile phones are connected to Raspberry Pi using the interface of the mobile application that we created. The processing tool is always listening in case of data received from Arduino, then data is stored in a database server. Then, data is examined and processed in a big data management platform in the Cloud. This process is repeated every 1 min. Our database contains 100,000 JSON documents of real-world healthcare data collected on real time from both mobile phones and sensors.

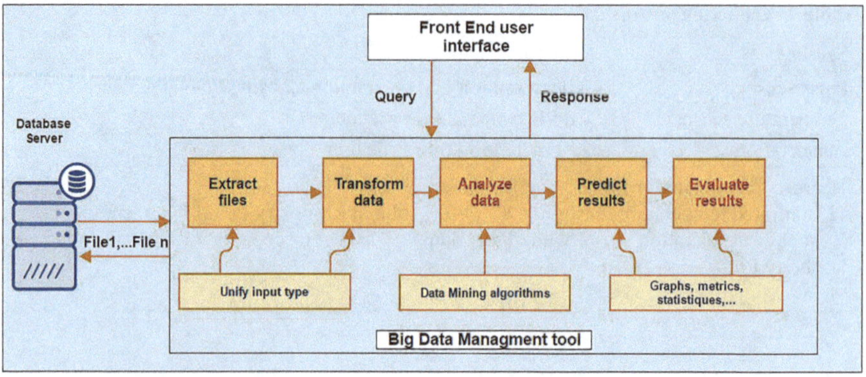

Fig. 4 Gathering sensors data workflow

3.2 Prediction Process

To get predicted results and transmitted to the end user, we follow the process in Fig. 4. The prediction process and the data collection are complementary because gathering inaccurate inputs or even having a weak prediction process will lead to incorrect results.

4 Case Study on Miscarriage Prediction

4.1 State of the Art of Miscarriage Prediction

The authors in [32] introduce the diagnosis of miscarriage by transvaginal ultrasound (TVS) assessment. But complete miscarriage diagnosis will not be complete without serial biochemical confirmation. According to this study, three-dimensional ultrasound has not improved in miscarriage prediction and diagnosis. Authors assert that the prediction of miscarriage using both ultrasound features and maternal history can be helpful in predicting pregnancy outcomes to react in advance.

In [12], the authors used ultrasound screening for the prediction of miscarriage in the first trimester. The study included 150 pregnant women within 5–12 weeks of gestation. Sixteen percent of women abort during the first trimester. Most of the feticides happen at 8–12 weeks of gestation. This study also addresses the importance of performing obstetric practice and learning obstetric ultrasound with efficiency and accuracy.

In a research study [13], the authors developed a new prediction system of miscarriage. The study shows that advancing maternal age increase the risk of having a subsequent miscarriage. While in women with a higher embryonic heart rate (HER) with an important volume of GS and crown rump, the risk of miscarriage is reduced. The new model was compared with the previous model developed in the same unit and evaluated using ROC curves.

In [14], the authors assert that BMI is a good risk factor during pregnancy. In fact, underweight (low BMI) and obese women (high BMI) have a higher probability of having a miscarriage. That explains the importance of weight and obesity during the stage of pregnancy.

In other research [15], alcohol consumption, maternal age, underweight, and obesity are named as the most significant risk factors for miscarriage. In fact, the probability of having miscarriage will decrease by 25.2% if we bring all these risk factors to low risk degree. On the other hand, another cohort study [8] identified crouching, physical activity, and sedentary work as risk factors for miscarriage in this cohort Shanghai textile worker.

The idea behind this study is to take benefit of technologies such as wearable sensors and mobile phones to predict miscarriage to react in advance in case of a probable miscarriage. We cannot deny that nowadays, those technologies made a revolution in understanding and analyzing human lifestyle due to the enormous data collected every second.

With respect to all works mentioned above, we propose a new predictive and preventive system that makes use of machine learning and big data tools; and it uses real-world data gathered from smartphone/sensors on real time. All works done in the literature related to miscarriage prediction used medical data and ultrasound screening for miscarriage diagnosis and not for prediction, since a miscarriage can happen before consultations and the pregnant woman lose her baby. The novel aspect of our research work is the use of real-time healthcare data gathered from both mobile phones and wearable sensors to monitor pregnancies by doctors and proactively give recommendations in advance to avoid a possible miscarriage during the whole period of pregnancy.

4.2 *Miscarriage Risk Factors*

Mobile phones and wearables are nowadays powerful tools to provide real-time and contextual information [17, 18]. In this study, we distinguish two sorts of data sources: data collected from healthcare sensors and data collected from mobile phones (see Table 2).

4.3 *System Dataset*

During our research work, we proposed an android application, namely, "e-Preg Monitoring." It assembles real-time data (from healthcare sensors and mobile phone) about the pregnant woman every 60s. Our database carries 100,000 documents gathered from 30 pregnant women from the period 2017/2018 and 15

Table 2 Miscarriage risk factor experiment

Source of data	Reference	Risk factor	Description
Healthcare sensors	[19, 20]	*Heart RATE Variability (HR)*	A marker of stress A high value of HR is a sign of a hypertension and high-rise blood pressure
	[19]	*Stress and blood pressure (BP)*	Values define based on the value of HRmax $HR_{max} = 208 - 0.7age$
	[21]	*Temperature variation (TP)*	A spontaneous miscarriage or premature delivery is associated with any viral infection A higher body temperature and flu increase the risk of having a miscarriage
	[16]	*Physical activity*	An supreme activity of the body is correlated with an elevated risk of miscarriage
Mobile phone	[10, 11]	*BMI*	A sign of obesity and underweight $BMI = \frac{W\,(kg)}{H^2\,(m)}$ Where W and H are, respectively, the weight and the height of the pregnant woman
	[6]	*Number of previous miscarriages*	Miscarriage is categorized by previous number previous pregnancy losses
	[8, 16]	*Maternal age*	Increased maternal is associated with and increased chance of having miscarriage
	[9]	*Location*	Food safety and eating well play well during pregnancy to avoid getting an illness
	[8]	*Current activity*	Activities like running, walking, biking, among others can encourage having miscarriage

valued attributes [22]. The file is available on the GitHub platform through the link "https://github.com/hibaasri/Miscarriage-Prediction". K-means clustering algorithm implemented in Machine Learning Library (MLlib) in Spark Databricks platform is used to make the predictive model [7, 27].

Some attributes of our dataset will not be represented during the prediction process; we cite: "type," "time," "ID," and "user_email." Hence, we have to recognize attributes, which will be included in *k*-means training for prediction, and

Table 3 Description of clusters

Cluster	Value	Prediction result
Cluster 1	0	No miscarriage
Cluster 2	1	Miscarriage

Table 4 Criteria of k-means algorithm

Parameters	Value (s)
Time to build model	0.51
Centers establishment	0.30
Cluster establishment	0.60
Model evaluation	0.19

to assemble them to a vector class to be implemented by *k*-means. To implement *k*-means in our research work [22]:

1. *Convert the data to a Resilient Distributed Datasets (RDD).*
2. *Train the model by passing the following parameters: the variable K (number of clusters), the number of iterations, and k-means specifications.*
3. *Create plots to visualize clusters.*
4. *Address the results to the doctors through the mobile application. We categorized this result through two conceivable values: "0" or "1" as presented in Table 3.*

4.4 Experiment Results

4.4.1 Performance and Efficiency

During the experience, we collect important information about the analysis and processing of K-means model. In this context, we deduce parameters like the time to build the model, cluster distribution, cluster definition, and model evaluation. Table 4 shows the values of those variables.

4.4.2 Clusters Definition and Distribution

Once we load all data, *k*-means sorts the data into two possible groups and builds the predictive model using representative attributes, namely: *age, BMI, nMisc, location, temp, BPM, stress,* and *BP*. To know the number of clusters, we define a parameter *K*, which represents the number of clusters, to the algorithm during the training; this choice will be, for sure, evaluated and validated later.

Besides, we communicate the results to the doctor through his profile in the mobile application. He can keep monitoring and tracking his patient to take quick decisions in case of a censorious case, while recommendations and advices are sent to the pregnant woman based on her behavior to not hurt her in a case of a probable miscarriage [26] (see Figs. 5 and 6).

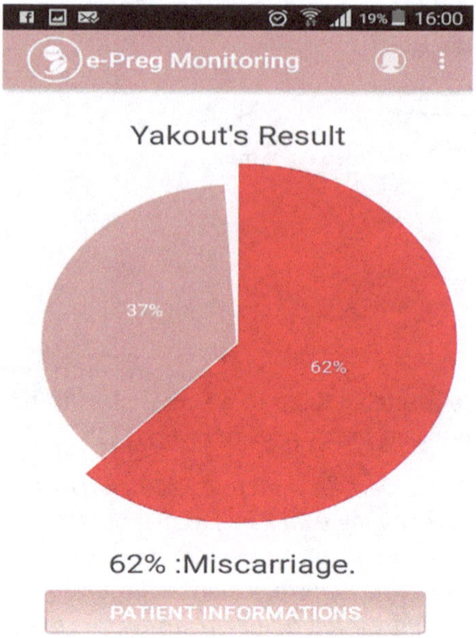

Fig. 5 Cluster distribution interface

Fig. 6 Recommandations interface

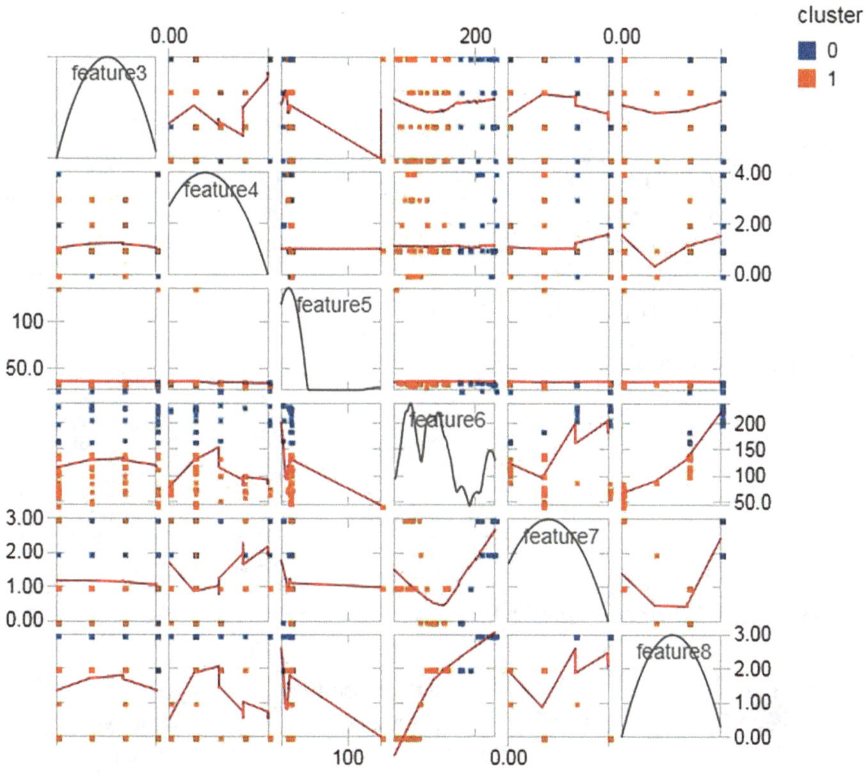

Fig. 7 *K*-means scatter plot

4.4.3 *K*-Means Scatter Plot

The scatter plot of k-means (see Fig. 7) shows that features "*F3*," "*F4*," "*F5*," "*F6*," "*F7*," and "*F8*" are the most representative attributes. Many reasons can explain this fact:

- During the day, the pregnant woman can do various activities.
- The pregnant woman can access different locations.
- She can get infections, so the value of her body temperature will change.
- Variations of heart rate and stress and blood pressure change.
- During pregnancy, the pregnant woman can lose or gain weight.

Table 5 constitutes a summary of variations and meaning of all attributes.

Table 5 Meaning and variation of attributes

	Features								
Summary	F0	F1	F2	F3	F 4	F5	F 6	F7	F 8
Value of	Age	BMI	nMisc	Activity	Location	Temp	BPM	Stress	BP
Mean[a]	25	19.84	2.0	2.51	1.44	36.0	128.04	1.21	1.51
Std dev[b]	0.0	0.0	0.0	1.11	1.07	1.25	58.08	1.14	1.11
Minimum[c]	25	19.84	2	1	0	28.0	43	1	1
Maximum[d]	25	19.84	2	4	4	40.0	238	3	3

[a]Average of feature's values
[b]Dispersion of the set of data values
[c]Small value of the feature
[d]High value of the feature

Table 6 Value of WSSSE's values for "$K = 1, 2, 3$"

Value of "K"	Value of WSSSE
1	2000.05
2	0978.24
3	0546.35

4.5 Evaluation and Validation

The main aim of k-means algorithm is to decrease the sum of squares of distance (the squared Euclidean distance) of points for each group. It is true that getting the prediction results is the goal, but the phase of evaluation represents an important task to validate the proposed model.

4.5.1 Evaluation Using a Random *K*

To evaluate the value of *K*, *k*-means proposes within set sum of squared errors (WSSSE) [23]. In our study, we choose "*K*" value of 2. In general, when "*K*" becomes higher, we get accurate results because we lessen the squared error. In our case, having a higher number of *K* is not representative because our goal is to predict the outcome of having a miscarriage or not. To evaluate this value, we train the model using different values of *K* (see Table 6).

From results above, we can notice that the appropriate value of "*K*" is 2 with a value of 978.24 in the WSSSE. We cannot choose 3 as the value of "*K*," because it is not our objective, and also choosing 1 as the value of "*K*" is not a clustering study. With 2 as the value of "*K*", we reduce error by 55%.

4.5.2 Clustering Validation Metrics

In clustering literature, various statistic parameters exist to evaluate and validate predictive models. We distinguish three main categories of those parameters [24, 25].

1. *Relative clustering validation (RCV).*
2. *Internal clustering validation (ICV).*
3. *External clustering validation (ECV).*

Relative Clustering Validation (RCV): Elbow Method

(i) *Elbow Method*

The principle behind this method is that we have to choose a number of clusters so that adding another cluster does not improve much better the total WSS (cluster's number). To determine the ideal number of clusters, we search for the location of a bend (knee) in the plot.

(ii) *Elbow Method's Results*

From the plot below (see Fig. 8), we can notice that the knee is located in the plot for 2 as the value of *K*. It becomes clear that our results based on choosing 2 as the value of "*K*" are reliable.

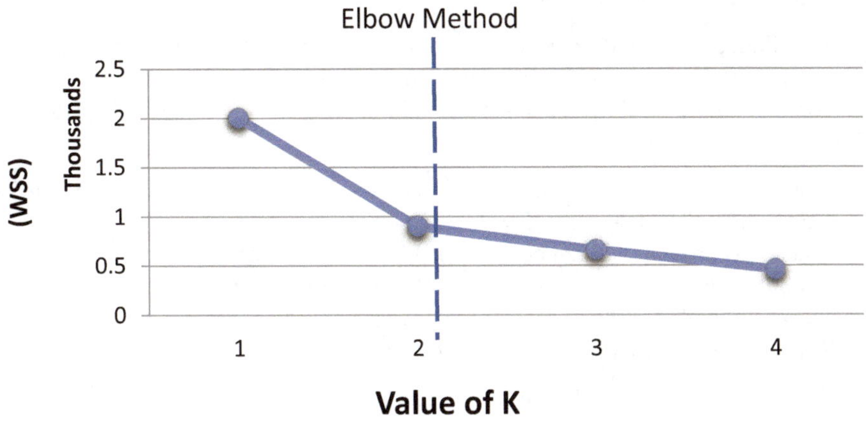

Fig. 8 Elbow method plot

```
val evaluator = new ClusteringEvaluator()
val silhouette = evaluator.evaluate(predictions)
```

▸ (2) Spark Jobs

```
silhouette: Double = 0.999999999947614
```

Fig. 9 Silhouette method

Internal Clustering Validation (ICV): Silhouette Method

(i) *Silhouette Method*

The silhouette method is a good indicator to determine if an observation is effectively clustered. The silhouette chart displays a measure of how close each point in one cluster is to points in the neighboring clusters. We determine the silhouette width S_i for each observation i, by:

$$S_0 = \frac{(b_i - a_i)}{\max(a_i, b_i)} \quad (Si \text{ ranges between } -1 \text{ and } 1)$$

Where:

$b_i = min_C\ d(i,C)$ *where C and d(i,C) represent, respectively, the other clusters and the average dissimilarity d(i,C) of i to all observations of C.*

(ii) *Silhouette Method's Results*

The silhouette method can be calculated through both of the following functions: "*fviz_silhouette*" or the function *fviz_nbclust*as (see Fig. 9).

To interpret the results obtained by the silhouette method, we have to analyze the value of *Si* found (see Fig. 9):

- When S_i is almost 1, we can deduce that observations are very well clustered. This is the case when using $K = 2$, we reach an $Si = 0.95$.
- When S_i is around 0, it can be explained by using over clusters (case of $K = 3$). Also, this value of S_i can signify that objects are not very well suitable to their cluster (case of $K = 1$).
- When S_i is negative, observations are now well situated.

From the plot in Fig. 10, we attend the highest width of 0.99 when $K = 2$. Therefore, we can conclude that points are well situated.

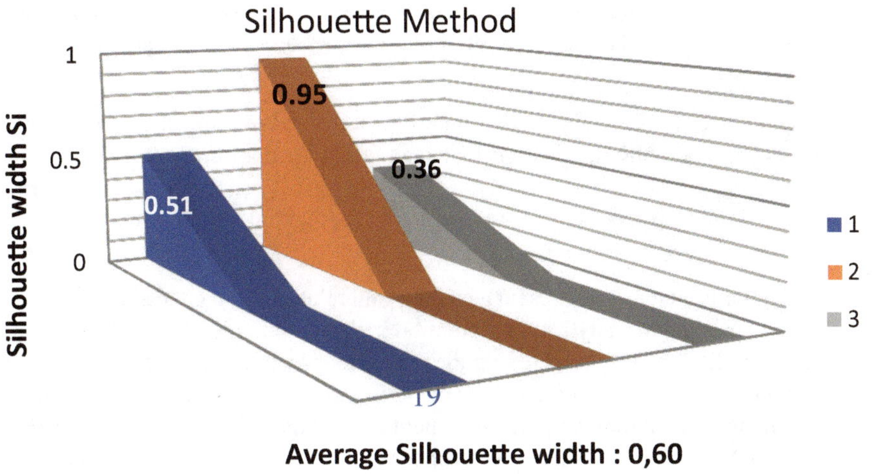

Fig. 10 Silhouette method plot

5 Conclusion

Big data and reality mining improve the healthcare system by enhancing medicine, reducing costs, understanding diseases, and making patient's life easier. The focus through this chapter is to take advantages and benefits of both reality mining and big data in order to predict outcomes in advance and involve patients in treatments. Big data improved our lives positively in many fields including scientific researches, administration, health care, economy, and business. Especially when dealing with live or death of the patient in healthcare system, big data faces many challenges including volume, veracity, velocity, variety of data, data value, data storage and capture, data transmission, data curation, data analysis, and data visualization.

By combining big data and reality mining, new performance solutions can be made to perform our lives, by detecting outcomes in advance, detecting healthcare fraud, and reducing costs. We can deduce the utility of machine learning and data mining in predicting patterns such as breast cancer, skin cancer, and obesity.

In this context, we propose an e-monitoring predictive and preventive system of disease. Through this book chapter, we presented the implementation of the system including data gathering process, the integration and the implementation of IoT systems, and the prediction process. The system described can be applied to any case study that uses data from mobile devices and healthcare sensors in the prediction process. This system benefits from machine learning and big data tools using real-world data gathered from smartphones/sensors on real time. Hence, we proposed an e-monitoring real-time miscarriage prediction system where both pregnant women and doctors are involved. We also designed a new mobile application "e-preg monitoring." The new application enables doctors monitoring his patients and recommends the pregnant women, during pregnancy's period, in order to avoid a probable miscarriage. The adoption of our system has many advantages such as:

- Reducing the number of miscarriages.
- Using real-time data in the prediction process, instead of ultrasound screening.
- Possessing a mobile application that interacts with both doctors and pregnant women.
- Having a nice experience of pregnancy (e.g., receiving positive recommendations).

Through this chapter, we present a background related to the good sides of reality mining and predictive analytics for prediction in different fields. We then present the proposed prediction system including the gathering process, the integration and the implementation of IoT systems, and the prediction process. The data conception, the big data management tool, and the implementation of the system are also described. To concrete the proposed system, we propose a case study of real-time miscarriage prediction that uses the propose approach. Through the section of the case study, we present the dataset created used, machine-learning algorithms, big data tools, and clustering evaluation/validation techniques.

References

1. Asri H, Mousannif H, Al Moatassime H, Noel T. Big data in healthcare: Challenges and opportunities. *2015 Int Conf Cloud Technol Appl.* 2015:1–7. doi:https://doi.org/10.1109/CloudTech.2015.7337020.
2. Mallu L, Ezhilarasie R. Live migration of virtual machines in cloud environment: A survey. *Indian J Sci Technol.* 2015;8(August):326–332. doi:https://doi.org/10.17485/ijst/2015/v8i.
3. Ngai, E. W., Xiu, L., & Chau, D. C. (2009). Application of data mining techniques in customer relationship management: A literature review and classification. *Expert systems with applications*, *36*(2), 2592–2602.
4. Yavaş, G., Katsaros, D., Ulusoy, Ö., & Manolopoulos, Y. (2005). A data mining approach for location prediction in mobile environments. *Data & Knowledge Engineering*, *54*(2), 121–146.
5. Kusiak, A., Zheng, H., & Song, Z. (2009). Short-term prediction of wind farm power: A data mining approach. *IEEE Transactions on energy conversion*, *24*(1), 125–136.
6. D. Bollier, The Promise and Peril of Big data. 2010, pp. 1–66.
7. Anon. n.d. "Apache Spark™ - Lightning-Fast Cluster Computing." Retrieved April 29, 2017 (http://spark.apache.org/).
8. Wong EY, Ray R, Gao DL, et al. Physical activity, physical exertion, and miscarriage risk in Women textile Workers in Shanghai, china. *Am J Ind Med.* 2010;53:497–505. doi:https://doi.org/10.1002/ajim.20812.
9. Women P. Food Safety for Pregnant Women. *Intern Med News*. 2005;38:79. doi:https://doi.org/10.1016/S1097-8690(05)72089-8.
10. Pasquali, R. (2006). Obesity, fat distribution and infertility. *Maturitas*, *54*(4), 363–371.
11. Toma C, Veleva Z, Tiitinen A, et al. High and low BMI increase the risk of miscarriage after IVF/ICSI and FET. 2008;23:878–884.
12. Kumari, Gynecology. 2016. "Prediction of Early Pregnancy Failure by Use of First Trimester Ultrasound Screening." *Int J Reprod Contracept Obstet Gynecol* 5(7):2135–40. Retrieved May 15, 2019 (doi:https://doi.org/10.18203/2320-1770.ijrcog20161897).

13. Stamatopoulos, Nicole et al. 2015. "Prediction of Subsequent Miscarriage Risk in Women Who Present with a Viable Pregnancy at the First Early Pregnancy Scan." *Australian and New Zealand Journal of Obstetrics and Gynaecology* 55(5):464–72. Retrieved May 15, 2019 (http://doi.wiley.com/10.1111/ajo.12395).
14. Why can't we conceive? | Family Doctor. http://www.familydoctor.co.uk/unassigned-articles/why-cant-we-conceive/. Accessed April 18, 2017.
15. Nizard J, Guettrot-Imbert G, Plu-Bureau G, et al. Pathologies maternelles chroniques et pertes de grossesse. Recommandations françaises. *J Gynecol Obstet Biol la Reprod.* 2014;43:865–882. doi:https://doi.org/10.1111/1471-0528.12694.
16. Wong, E. Y., Ray, R., Gao, D. L., Wernli, K. J., Li, W., Fitzgibbons, E. D., … & Thomas, D. B. (2010). Physical activity, physical exertion, and miscarriage risk in women textile workers in Shanghai, China, American journal of industrial medicine, 53(5), 497–505.
17. Dimpfel F. a Smart, Location Based Time and Attendance Tracking System Using Android Application. *Integr Vlsi J.* 2001;5(1):1–22. doi:https://doi.org/10.5121/ijcseit.2015.5101.
18. Asri, H., Mousannif, H., & Al Moatassime, H. (2017). Real-time Miscarriage Prediction with SPARK. Procedia Computer Science, 113, 423–428.
19. Thayer JF, Åhs F, Fredrikson M, Sollers JJ, Wager TD. A meta-analysis of heart rate variability and neuroimaging studies: Implications for heart rate variability as a marker of stress and health. *Neurosci Biobehav Rev.* 2012*;36(2):747–756.* doi:https://doi.org/10.1016/j.neubiorev.2011.11.009.
20. Tanaka H, Monhah K, Seals D. Age-predicted maximum heart rate revisited. *J Am Coll Cardiol. 2001;37(1):153–156.*
21. Anselem O, Floret D, Tsatsaris V, Goffinet F, Launay O. Influenza infection and pregnancy [French; English] Grippe au cours de la grossesse. *Press Medicale.* 2013;42(11):1453–1460. doi:https://doi.org/10.1016/j.lpm.2013.01.064.
22. Asri, H., Mousannif, H., & Al Moatassime, H. (2018). Comprehensive miscarriage dataset for an early miscarriage prediction. Data in Brief, 19, 240–243.
23. Chai T, Draxler RR. Root mean square error (RMSE) or mean absolute error (MAE)? Arguments against avoiding RMSE in the literature. *Geosci Model Dev.* 2014;7(2005):1247–1250. doi:https://doi.org/10.5194/gmd-7-1247-2014.
24. Theodoridis, S., & Koutroumbas, K. (2008). Pattern Recognition & Matlab Intro. Pattern Recognition.
25. Charrad, M., Ghazzali, N., Boiteau, V., Niknafs, A., & Charrad, M. M. (2014). Package 'nbclust'. Journal of statistical software, 61, 1–36.
26. Asri, H., Mousannif, H., & Al Moatassime, H. (2019). Reality mining and predictive analytics for building smart applications. *Journal of Big Data, 6*(1), 1–25.
27. Asri, H., Mousannif, H., & Al Moatassime, H. (2019). Big data analytics in healthcare: case study-miscarriage prediction. *International Journal of Distributed Systems and Technologies (IJDST), 10*(4), 45–58.
28. Asri, H., Mousannif, H., Al Moatassime, H., & Noel, T. (2016). Using machine learning algorithms for breast cancer risk prediction and diagnosis. *Procedia Computer Science, 83,* 1064–1069.
29. Asri, H., Mousannif, H., & Al Moatassim, H. (2019, July). A Hybrid Data Mining Classifier for Breast Cancer Prediction. In *International Conference on Advanced Intelligent Systems for Sustainable Development* (pp. 9–16). Springer, Cham.
30. Mousannif H., Asri H., Mansoura M., Mourahhib A., Isaouy Y., Marmouchi M. (2021) Skin Cancer Prediction and Diagnosis Using Convolutional Neural Network (CNN) Deep Learning Algorithm. In: Ben Ahmed M., Rakıp Karaş İ., Santos D., Sergeyeva O., Boudhir A.A. (eds) Innovations in Smart Cities Applications Volume 4. SCA 2020. Lecture Notes in Networks and Systems, vol 183. Springer, Cham. doi:https://doi.org/10.1007/978-3-030-66840-2_42
31. Asri, H., Mousannif, H., Al Moatassime, H., & Zahir, J. (2020, June). Big Data and Reality Mining in Healthcare: Promise and Potential. In *International Conference on Image and Signal Processing* (pp. 122–129). Springer, Cham.

32. Coomarasamy, A., Devall, A. J., Brosens, J. J., Quenby, S., Stephenson, M. D., Sierra, S., . . . & Gallos, I. D. (2020). Micronized vaginal progesterone to prevent miscarriage: a critical evaluation of randomized evidence. *American journal of obstetrics and gynecology*.
33. Wood, D. A. (2021). Net ecosystem carbon exchange prediction and insightful data mining with an optimized data-matching algorithm. *Ecological Indicators*, *124*, 107426.

Smart System for Monitoring and Controlling of Agricultural Production by the IoT

Jamal Mabrouki, Karima Azoulay, Saloua Elfanssi, Loubna Bouhachlaf, Fatimazahra Mousli, Mourade Azrour, and Souad El Hajjaji

1 Introduction

Agriculture refers to human activity that extremely acts in natural environments in order to produce some goods useful in human life, including food, drinking, clothes, furniture, etc. It occupies a significant proportion of the land surface, with major consequences to be useful to humans. The impacts of agriculture on soils and biodiversity have existed since its emergence, about 10,000 years ago. The environmental impacts of contemporary agriculture extend beyond agricultural ecosystems and include water and air pollution, contribution to global warming.

In accordance with the International Telecommunication Union, the Internet of Things (IoT) can be denoted as global interface for the information community. IoT makes advanced practices possible by connecting tangible or intangible objects through interactive information and telecommunication systems. In fact, the notion of the Internet of Things is not definitively determined [1–3]. It covers both theoretical and practical aspects. Conceptually, the Internet of Things refers to physically connected objects that have their own digital identification and can communicate with each other. Technically, the Internet of Things corresponds to

J. Mabrouki (✉) · K. Azoulay · L. Bouhachlaf · F. Mousli · S. El Hajjaji
Faculty of Science, Laboratory of Spectroscopy, Molecular Modeling, Materials, Nanomaterial, Water and Environment, CERNE2D, Mohammed V University, Rabat, Morocco
e-mail: jamal.mabrouki@um5r.ac.ma

S. Elfanssi
National Center for Research and Studies on Water and Energy (CNEREE), Cadi Ayyad University, Marrakech, Morocco

M. Azrour
Department of Computer Science, Faculty of Sciences & Techniques, Moulay Ismail University, Errachidia, Morocco

M. Azrour et al. (eds.), *IoT and Smart Devices for Sustainable Environment*, EAI/Springer Innovations in Communication and Computing,
https://doi.org/10.1007/978-3-030-90083-0_8

a standardized and integrated digital identification system of IP address, protocols like SMTP, and HTTP; these objects take advantage of wireless networks including RFID, Bluetooth, or Wi-Fi [4–7].

On the other side, recent technological developments have influenced all daily activities including transportation, healthcare, surveillance, water monitoring, electricity grid, and environment [8–17]. Hence, agricultural production is also being affected. Through the employment of numerous wireless sensor devices and densitometry devices, researchers can gather information about the moisture content of the soil [18], the condition of fertilizers in the soil [19], or the risk of plant disease spread and the presence of insect pests [20].

The employment of sensors to gather and collect data on the work of agricultural systems enables the persons in charge to modulate the operation of the devices according to the real situations, in order to achieve an improvement in the efficiency and quality of the production process [21]. The Internet of Things has proposed a new production perspective to conventional agriculture: the information system in the form of a network with which it is associated and necessitates automatization, the presence of intelligent devices, and their integration into the agricultural production process. Consequently, farmers are able to interact in real time and propose effective solutions.

The earliest application of agricultural automation technology was in the 1980s; it begins with the development of machines designed to process milking cows and collect and gather fruits. At that time, the objective was basically economic: it was important to minimize the production-related costs [22]. On the contrary, the effects of the implementation of the Internet of Things in today's agricultural sector have particular benefits that include increased effectiveness in the usage of the land, the water, the fertilizers, the pesticides, and the other inputs, the reduction of the production costs, the enhancement of the profitability, the sustainability, the food safety, and the protection of the environment [23–25].

This current research is dedicated to the introduction of the Internet of Things concept in the context of a monitoring system of designs implemented in agricultural production. In addition, the research also focuses on the specification of IoT devices and their implementation via the Arduino board-based platform.

The rest of this paper is organized as follows: In Sect. 2, we present some related works. In Sect. 3, we detail the used materials. Section 4 is reserved for specifying our proposed system. The results and discussion are described in Sect. 5. Finally, the paper concludes in the last section.

2 Related Works

In this section, we discuss some related works proposed in the literature for the improvement of the agriculture sector. The enthusiasm of many scientists has produced the design of enhancement modes and models for the development of the irrigation system, hence economizing the used water quantity. In addition, some researchers have demonstrated that agricultural statistical data and statistical data on insects, liquid consumption, and meteorites are beneficial for better exploitation of pesticide resources.

In 2016, Nikesh Gondchawar et al. [26] published a study on IoT-based smart agriculture. This research attempts to design smart agriculture through the application of both automation and IoT-based technologies. The designed system involves smart irrigation and smart control. Furthermore, it can take smart decisions based on accumulated real-time data in the field. The system is able to monitor temperature, humidity, and theft detection in the storage building. The whole things are controlled by a smart device and are accomplished by interfacing sensors, Zigbee modules, camera, and actuators that are connected to Raspberry Pi. In the same year, Tanmay Baranwal et al. [27] address the issue of protecting agricultural crops from damage caused by pests or insects in the field or in grain storage. Monitoring mechanisms are implemented to report identified problems in real time. Sensors and technical equipment are combined using Python codes. Besides, the system operating based on an algorithm that is developed for gathering data and allowing an exact alert of the user.

In 2017, Prathibha et al. [28] have proposed a new intelligent farming system. The designed tool is more powerful and more advantageous for the farmers. It reports the air temperature and humidity measured in agricultural fields to the farmer through MMS message. In addition, the new device can be implemented in greenhouses and in thermal-sensitive plants.

In 2019, Muangprathub et al. [29] have developed a solution to optimally control the irrigation of agricultural plants through a wireless sensor network. The goal of this work was to create and develop a control system using sensors and management of irrigation via a smartphone and a web application.

In 2020, Lavanya et al. [30] presented a system based on the Internet of Things (IoT). The authors developed a completely novel nitrogen–phosphorus–potassium (NPK) sensor by using light-dependent resistor (LDR) and light-emitting diodes (LED). The colorimetric method is employed to supervise and evaluate the existent nutritive elements in the field. The results collected by the NPK sensor in the chosen agricultural sites are transmitted to the Google Cloud database.

In 2021, Pengwen Wang et al. [31] proposed an upgraded multilayer perceptron method for predicting the quantity of sugar revenue production in IoT agriculture. Then, the authors demonstrated that their proposition, which is based on MLP perceptron artificial neural network counting two hidden layers, is further effective when compared with the four known classifier algorithms, which include support vector machine, random forest, decision tree, linear regression, and naïve base.

3 System Development Materials

As depicted in Fig. 1, our proposed system consists of five main modules that are: sensing module, actuator module, communication, processing module, and

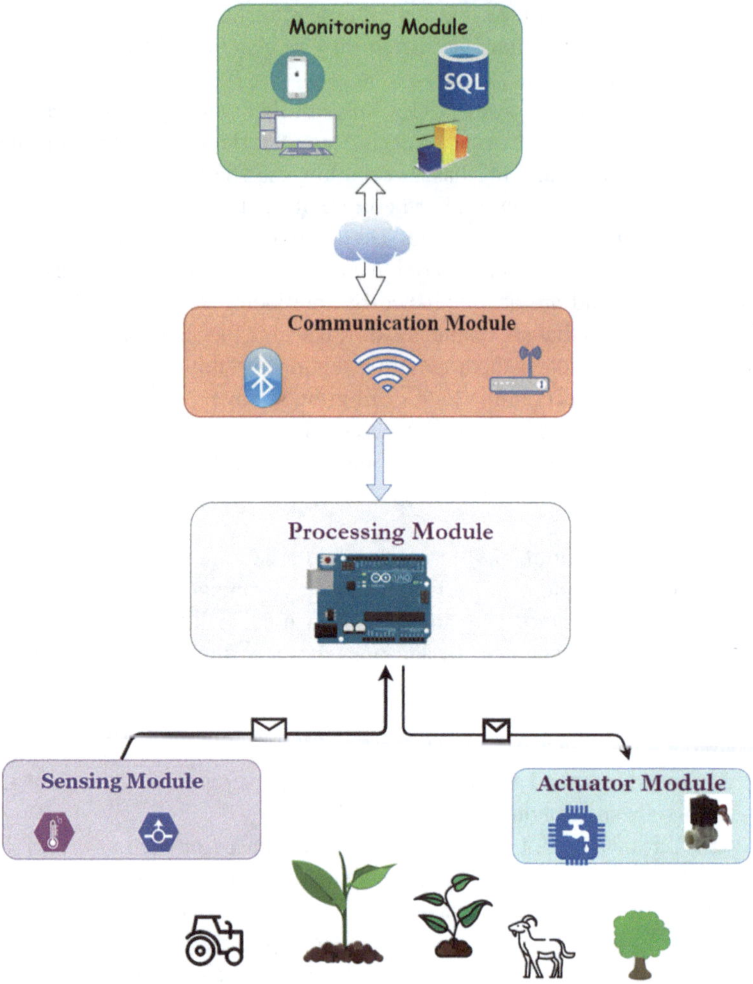

Fig. 1 Architecture of designed system

monitoring module. So, in this section, we explain the role of each module and its relationship with others. Then, we give information about the used materials.

3.1 Processing Module

In this module, we have adopted Arduino board [32]. This board system is designed from an Open Source platform installed on an AVR microcontroller programmed board allowing the writing, compiling, and testing of a program. Arduino boards

Table 1 Different used sensors

N°	Sensor	Parameter to measure	Measurement unit
1	DHT11	Air temperature	°C (Celsius degree)
		Air humidity	Take values (0–100%)
2	LDR	Light intensity	Take values (0–100%)
3	Water flow	Water flow rate	L/m (letter/minute)
4	Soil moisture sensor	Soil humidity	Take values (0–100%)

and modules are designed with inputs/outputs that receive signals from sensors or control switching devices. For example, in a smart home [33], Arduino can switch On/Off the light of a room after receiving the signal from sensor light. The Arduino programs are controlled or can be controlled via a computer. The Arduino programming contains a wiring implementation language. The development platform is based on a multimedia sphere of programming processing [34].

3.2 Sensing Module

Actually, sensors are small tools that can capture and measure environmental parameters [35–37]. Therefore, they are the source of data in IoT systems. In our system, various sensors were adopted for sensing specific values such as temperature, humidity, light, soil moisture, and water flow (see Table 1). The details about the used sensors are given in the following.

3.2.1 Temperature and Humidity Sensors

The DHT22 [38] and DHT11 [39] are temperature and humidity sensors. The DHT22/AM2302 device is capable of measuring air temperature between −40 and +125 °C with an accuracy of +/−0.5 °C and relative humidity levels from 0% to 100% with an accuracy of +/−2%. The temperature value can be measured in each ½ s. On the other hand, the DHT11 detector is able to measure temperature in the interval of 0 to +50 °C with the accuracy of +/−2 °C. Furthermore, it can measure relative humidity levels if it is between 20% and 80% with an accuracy of +/−5%. The DHT22 and the DHT11 are both compatible 3.3 and 5 V. They also have the same wiring and communication protocol. We notice here that there are other temperature and humidity sensors [40–42].

3.2.2 Light Sensor

Generally, agriculture is affected by weather conditions. Accordingly, in our project, we have to detect if it is sunny, cloudy, or foggy. To do that, we use light sensors

Fig. 2 Light sensor

that can detect the degree of received light energy in the field. The light sensor (Fig. 2) is a form of passive equipment that enables the transformation of both visible and invisible light into an electrical current signal. Due to their ability to transform light energy to electric power, they are frequently described as "photoelectric devices" or "photo sensors" [43].

3.2.3 Water Flow Sensor

The water flow sensor [44] consists of a plastic valve body, a water rotor, and a Hall effect sensor. When water flows through the rotor, it rolls, then its velocity changes according to the flow rate. This type of sensor can be found on different diameters, water pressure (MPa), and flow ranges (L/m). So, we have to pick the one that responds to our wishes.

3.2.4 Soil Moisture

The soil moisture sensor consists of two types of connectors that determine the degree of soil moisture. The two electrodes allow the electrical signal to transit through the soil and, based on its resistance capacity, determine the percentage of soil moisture. If the soil is wet, it is more conductive to electricity, which means that the resistance will be lower. The percentage of moisture will be higher. Arid soil decreases the conductivity capacity. Therefore, when there is less water, the soil is less electrically conductive, which means a higher resistance. Thus, the moisture level is lower [45].

3.3 Actuator Module

The developed system provides automatic irrigation of plants. In fact, it is based on the values gathered by the sensors of soil moisture, air temperature, and air humidity. These values are then transmitted to a programmable Arduino board. After analyzing the data, the controller sends commands to activate or deactivate the solenoid valve. In case of detection that the soil was too dry, the controller ordered the solenoid valve to open and irrigate the crops, whereas in the case of wet soil, the solenoid valve will turn off and the water will not flow down.

3.4 Communication Module

3.4.1 Bluetooth

For connecting our system to a local device, we adopt the Bluetooth module. This last one allows establishing a Bluetooth serial link between the Arduino card and numerous **devices** equipped with Bluetooth technology, including smartphone, tablet, laptop, and other ones. We notice at this point that there are two types of Bluetooth modules: HC-06 module and HC-05 module. Originally, the HC-06 module is “slave” modules, while the HC-05 module is a “master.” A “master” module can ask another Bluetooth device to pair with it while a “slave” one can only receive pairing requests [46].

3.4.2 Wi-Fi

In order to give our system the possibility to be connected to the Internet, we have used the Arduino Uno Wi-Fi module. The card is actually designed on the basis of the ATmega328P with an embedded ESP8266Wi-Fi unit. The ESP8266Wi-Fi module is an autonomous built-in circuit with an embedded TCP/IP protocol layer. This protocol allows to the device to access Wi-Fi network or give it the ability to act as an access point [47].

4 The Proposed System

4.1 System Overview

The present study is devoted to the use of various sensors for controlling and monitoring the environmental conditions of a field. The soil moisture sensor is applied to define the time when the plants require water. In fact, a soil moisture

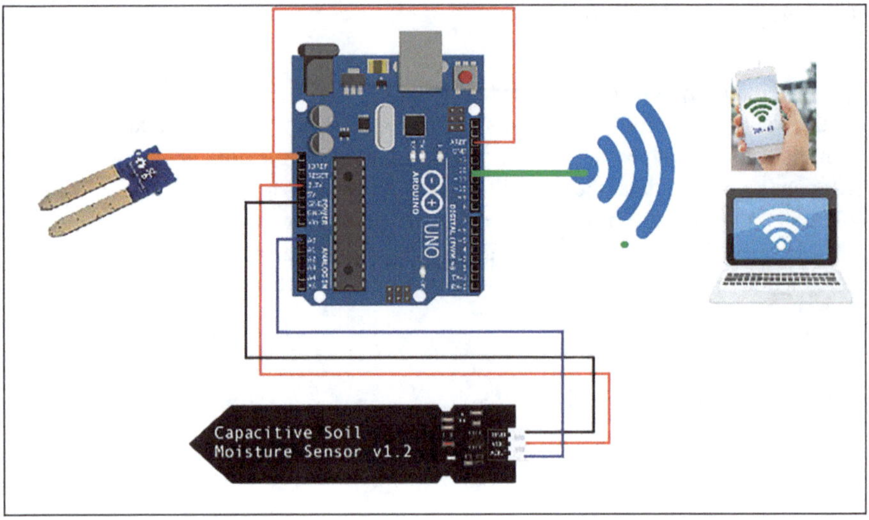

Fig. 3 Proposed system overview

sensor is able to measure the soil resistance. By the way, measurement indicates the soil moisture value.

According to Fig. 3, the modeling of the envisaged IoT system has been implemented using the Arduino board and appropriate sensors. The basic structure of the IoT node consists of an Arduino Uno board on which several connectors are mounted. Sensors are designed to collect the value of temperature and relative humidity, soil moisture, water flow, and light intensity. The parameters collected by these sensors connected to the analog and digital pins of the Arduino can be used, processed, or displayed locally or transferred to a remote server. Thanks to its connection to the Wi-Fi module, the Arduino board can send data, taking advantage of HTTP protocol. In fact, in this case, the IoT device acts as a web client, and the moment the sensor gets data, it can issue an HTTP request, which transmits the data to a web server on the Internet.

This monitoring data can then be analyzed to identify the crops that best meet the productivity objectives of any farm in the world. This irrigation method is the operation of artificially providing water to cultivated plants to help them grow and develop normally in the absence of water induced by rainfall deficits, drainage, or groundwater depletion, particularly in arid areas.

4.2 System Implementation

As explained above, Internet of Things devices are connected to the network by way of a Wi-Fi module router and are able to communicate data to a remote server via

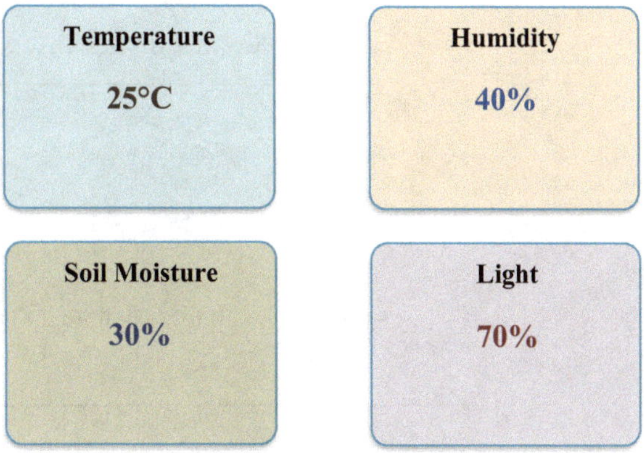

Fig. 4 Dashboard of web application

the Internet. The data collected from the sensors by the IoT devices are stored in a database that users can access. Those users can have access to such information from any location using a computer desktop, tablet, or smartphone. In addition to the system monitoring role, user tools also allow control of actuators via IoT devices over the Internet.

In order to give the administrator the manual ability to manage the different data in real time, coming from the IoT devices, we have developed web application. Thanks to this application, a supervisor may adjust the settings of the water consumption requirements of every crop. In particular, as shown in Fig. 4, the different details of the information coming from the IoT devices can be displayed by an administrator.

In addition, we have set up a lightweight program in Arduino board. This program can send commends to the electronic valve if the plants need water. The algorithm is depicted in Fig. 5.

5 Results and Discussion

In fact, as soon as we install the sensors on the ground, they take the values and transmit them to the web application. The interval between one measurement and another is 60 minutes. Eventually, the received values are displayed on a web page. We notice that the temperature and humidity variations have not changed in a big range, as we can see on the graph (Fig. 6). These values are used in order to take an efficient decision in order not to lose the cultivations and also to economize water. The temperature, the humidity of the soil, and the rate of light are very important factors for any type of agriculture. For example, in case the temperature value is between 0 and 10 °C, the rate of photosynthesis will decrease enormously. The same things can happen if it exceeds 40 °C, but in this case, the enzymes, which perform the photosynthesis, will lose their form.

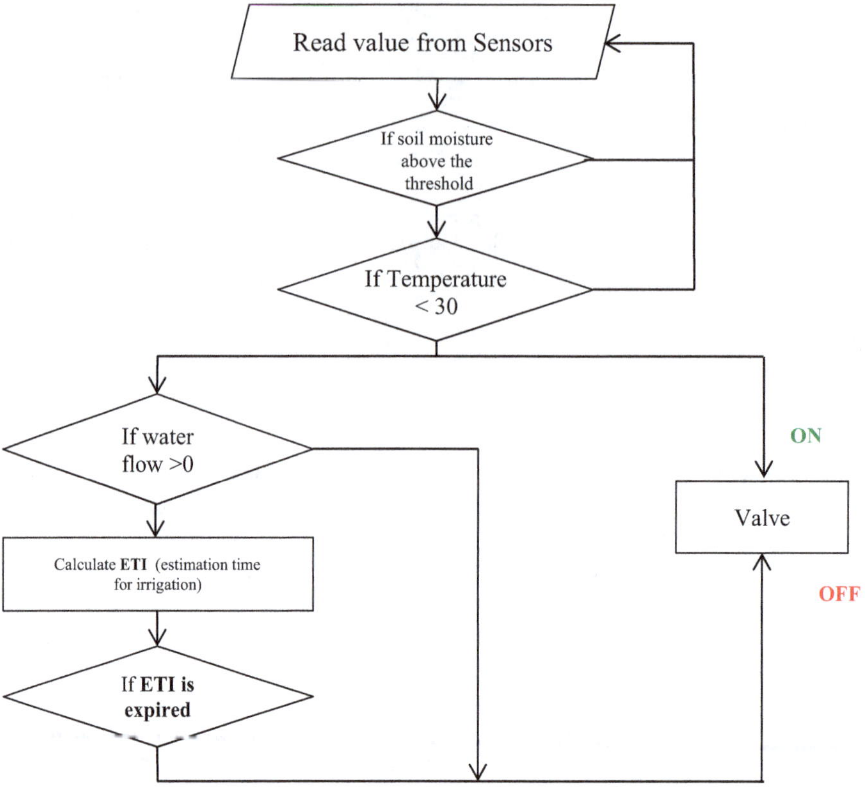

Fig. 5 Programmed algorithm

6 Conclusion

The IoT-based monitoring and control system concept presented in the present paper can be used on agricultural farms. The autonomous sensors play a key role in the gathering of data for the monitoring system, and they also send the signals to the actuators in the control process. Our overall goal was to develop a platform for smart greenhouse control with the ability to monitor and perceive environmental soil and weather situations in agricultural areas. The problem we tackled in the project was the automatic recognition of situations. In addition, the IoT concept was also introduced to connect the devices via the Internet and facilitate user access to information. IoT-based devices can also connect to and access physical objects in the server farm via the Internet, allowing users to remotely monitor production conditions and processes.

Fig. 6 Measurement of agricultural soil parameters

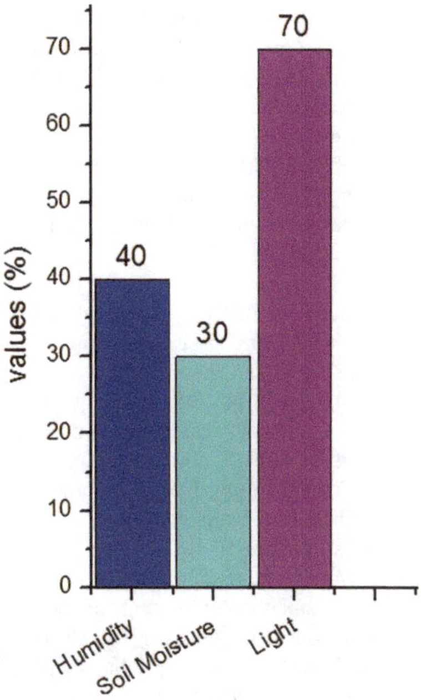

References

1. L. Atzori, A. Iera, and G. Morabito, "The internet of things: A survey," *Computer networks*, vol. 54, no. 15, pp. 2787–2805, 2010.
2. F. Wortmann and K. Flüchter, "Internet of things," *Business & Information Systems Engineering*, vol. 57, no. 3, pp. 221–224, 2015.
3. L. Srivastava and T. Kelly, "The internet of things," *International Telecommunication Union, Tech. Rep*, vol. 7, 2005.
4. B. N. Silva, M. Khan, and K. Han, "Internet of things: A comprehensive review of enabling technologies, architecture, and challenges," *IETE Technical review*, vol. 35, no. 2, pp. 205–220, 2018.
5. Y. Liao, E. de F. R. Loures, and F. Deschamps, "Industrial Internet of Things: A systematic literature review and insights," *IEEE Internet of Things Journal*, vol. 5, no. 6, pp. 4515–4525, 2018.
6. M. Wu, T.-J. Lu, F.-Y. Ling, J. Sun, and H.-Y. Du, "Research on the architecture of Internet of Things," in *2010 3rd international conference on advanced computer theory and engineering (ICACTE)*, 2010, vol. 5, pp. V5-484–V5-487.
7. F. Wortmann and K. Flüchter, "Internet of things," *Business & Information Systems Engineering*, vol. 57, no. 3, pp. 221–224, 2015.
8. J. Mabrouki, M. Azrour, G. Fattah, D. Dhiba, and S. E. Hajjaji, "Intelligent monitoring system for biogas detection based on the Internet of Things: Mohammedia, Morocco city landfill case," *Big Data Min. Anal.*, vol. 4, no. 1, pp. 10–17, Mar. 2021, doi: https://doi.org/10.26599/BDMA.2020.9020017.
9. J. Mabrouki, M. Azrour, D. Dhiba, Y. Farhaoui, and S. E. Hajjaji, "IoT-based data logger for weather monitoring using Arduino-based wireless sensor networks with remote graphical application and alerts," *Big Data Min. Anal.*, vol. 4, no. 1, pp. 25–32, Mar. 2021, doi: https://doi.org/10.26599/BDMA.2020.9020018.

10. J. Mabrouki, M. Azrour, and S. El, "Use of Internet of Things for Monitoring and Evaluation water's Quality: Comparative Study," *International Journal of Cloud Computing*, 2021.
11. J. Mabrouki, M. Azrour, Y. Farhaoui, and S. El Hajjaji, "Intelligent System for Monitoring and Detecting Water Quality," in *Big Data and Networks Technologies*, vol. 81, Y. Farhaoui, Ed. Cham: Springer International Publishing, 2020, pp. 172–182.
12. H. Kim, H. Choi, H. Kang, J. An, S. Yeom, and T. Hong, "A systematic review of the smart energy conservation system: From smart homes to sustainable smart cities," *Renewable and Sustainable Energy Reviews*, vol. 140, p. 110755, Apr. 2021, doi: https://doi.org/10.1016/j.rser.2021.110755.
13. M. S. Aliero, K. N. Qureshi, M. F. Pasha, and G. Jeon, "Smart Home Energy Management Systems in Internet of Things networks for green cities demands and services," *Environmental Technology & Innovation*, vol. 22, p. 101443, May 2021, doi: https://doi.org/10.1016/j.eti.2021.101443.
14. Y. Mehmood, F. Ahmad, I. Yaqoob, A. Adnane, M. Imran, and S. Guizani, "Internet-of-Things-Based Smart Cities: Recent Advances and Challenges," *IEEE Commun. Mag.*, vol. 55, no. 9, pp. 16–24, 2017, doi: https://doi.org/10.1109/MCOM.2017.1600514.
15. S. Tian, W. Yang, J. M. L. Grange, P. Wang, W. Huang, and Z. Ye, "Smart healthcare: making medical care more intelligent," *Global Health Journal*, vol. 3, no. 3, pp. 62–65, Sep. 2019, doi: https://doi.org/10.1016/j.glohj.2019.07.001.
16. Z. Mohd Yusoff, Z. Muhammad, M. S. I. Mohd Razi, N. F. Razali, and M. H. C. Hashim, "IOT-Based smart street lighting enhances energy conservation," *IJEECS*, vol. 20, no. 1, p. 528, Oct. 2020, doi: https://doi.org/10.11591/ijeecs.v20.i1.pp528-536.
17. M. Sree Devi and V. Rahamathulla, "Air Quality Through IoT and Big Data Analytics," in *Advances in Data Science and Management*, Singapore, 2020, pp. 181–187, doi: https://doi.org/10.1007/978-981-15-0978-0_17.
18. M. Zhang, M. Li, W. Wang, C. Liu, and H. Gao, "Temporal and spatial variability of soil moisture based on WSN," *Mathematical and Computer Modelling*, vol. 58, no. 3–4, pp. 826–833, 2013.
19. S. G. Galande, G. H. Agrawal, and S. B. Dighe, "Greenhouse Microclimatic Real-Time Monitoring with the Help of NPK Sensor," *International Journal of Emerging Trends in Science and Technology*, vol. 2, no. 5, pp. 2511–2515, 2015.
20. A. E. Douglas, "Strategies for enhanced crop resistance to insect pests," *Annual review of plant biology*, vol. 69, pp. 637–660, 2018.
21. H. M. Jawad, R. Nordin, S. K. Gharghan, A. M. Jawad, and M. Ismail, "Energy-efficient wireless sensor networks for precision agriculture: A review," *Sensors*, vol. 17, no. 8, p. 1781, 2017.
22. T.-K. Dao, T.-S. Pan, and J.-S. Pan, "A multi-objective optimal mobile robot path planning based on whale optimization algorithm," in *2016 IEEE 13th International Conference on Signal Processing (ICSP)*, 2016, pp. 337–342.
23. T. S. T. Bhavani and S. Begum, "Agriculture productivity enhancement system using IOT," *International Journal of Theoretical and Applied Mechanics*, vol. 12, no. 3, pp. 543–554, 2017.
24. A. Salam and S. Shah, "Internet of things in smart agriculture: Enabling technologies," in *2019 IEEE 5th World Forum on Internet of Things (WF-IoT)*, 2019, pp. 692–695.
25. A. Khanna and S. Kaur, "Evolution of Internet of Things (IoT) and its significant impact in the field of Precision Agriculture," *Computers and electronics in agriculture*, vol. 157, pp. 218–231, 2019.
26. N. Gondchawar and R. S. Kawitkar, "IoT based smart agriculture," *International Journal of advanced research in Computer and Communication Engineering*, vol. 5, no. 6, pp. 838–842, 2016.
27. T. Baranwal, Nitika, and P. K. Pateriya, "Development of IoT based smart security and monitoring devices for agriculture," in *2016 6th International Conference - Cloud System and Big Data Engineering (Confluence)*, Jan. 2016, pp. 597–602, doi: https://doi.org/10.1109/CONFLUENCE.2016.7508189.

28. S. R. Prathibha, A. Hongal, and M. P. Jyothi, "IOT Based Monitoring System in Smart Agriculture," in *2017 International Conference on Recent Advances in Electronics and Communication Technology (ICRAECT)*, Bangalore, India, Mar. 2017, pp. 81–84, doi: https://doi.org/10.1109/ICRAECT.2017.52.
29. J. Muangprathub, N. Boonnam, S. Kajornkasirat, N. Lekbangpong, A. Wanichsombat, and P. Nillaor, "IoT and agriculture data analysis for smart farm," *Computers and Electronics in Agriculture*, vol. 156, pp. 467–474, Jan. 2019, doi: https://doi.org/10.1016/j.compag.2018.12.011.
30. G. Lavanya, R. C, and G. P, "An automated low cost IoT based Fertilizer Intimation System for smart agriculture," *Sustainable Computing: Informatics and Systems*, vol. 28, p. 100300, Dec. 2020, doi: https://doi.org/10.1016/j.suscom.2019.01.002.
31. P. Wang, "An Improved Multilayer Perceptron Approach for Detecting Sugarcane Yield Production in IoT based Smart Agriculture," *Microprocessors and Microsystems*, p. 7, 2021.
32. M. McRoberts, *Beginning Arduino*. Apress, 2011.
33. Q. I. Sarhan, "Arduino Based Smart Home Warning System," in *2020 IEEE 6th International Conference on Control Science and Systems Engineering (ICCSSE)*, 2020, pp. 201–206.
34. A. A. Galadima, "Arduino as a learning tool," in *2014 11th International Conference on Electronics, Computer and Computation (ICECCO)*, 2014, pp. 1–4.
35. M. Srbinovska, C. Gavrovski, V. Dimcev, A. Krkoleva, and V. Borozan, "Environmental parameters monitoring in precision agriculture using wireless sensor networks," *Journal of cleaner production*, vol. 88, pp. 297–307, 2015.
36. Z. Liu, C. Sun, and L. Xiang, "The measure of environmental sensitivity in detection performance degradation," in *AIP Conference Proceedings*, 2012, vol. 1495, no. 1, pp. 659–666.
37. G. Pavlogeorgatos, "Environmental parameters in museums," *Building and Environment*, vol. 38, no. 12, pp. 1457–1462, 2003.
38. M. Bogdan, "How to use the DHT22 sensor for measuring temperature and humidity with the Arduino board," *ACTA Universitatis Cibiniensis*, vol. 68, no. 1, pp. 22–25, 2016.
39. H. A. N. Dan-ao and W. Fei, "Research application of the digital temperature and humidity sensor DHT11 [J]," *Electronic Design Engineering*, vol. 13, p. 027, 2013.
40. R. Soukup, A. Hamacek, L. Mracek, and J. Reboun, "Textile based temperature and humidity sensor elements for healthcare applications," in *Proceedings of the 2014 37th international spring seminar on electronics technology*, 2014, pp. 407–411.
41. C. Zhang, W. Zhang, D. J. Webb, and G.-D. Peng, "Optical fibre temperature and humidity sensor," *Electronics Letters*, vol. 46, no. 9, pp. 643–644, 2010.
42. C.-Y. Lee and G.-B. Lee, "Micromachine-based humidity sensors with integrated temperature sensors for signal drift compensation," *Journal of micromechanics and microengineering*, vol. 13, no. 5, p. 620, 2003.
43. T. Dinh Le and D. H. Tan, "Design and deploy a wireless sensor network for precision agriculture," in *2015 2nd National Foundation for Science and Technology Development Conference on Information and Computer Science (NICS)*, Sep. 2015, pp. 294–299, doi: https://doi.org/10.1109/NICS.2015.7302210.
44. M. A. S. B. Azmi and S. Sulaiman, "Design of domestic water meter using Arduino," in *AIP Conference Proceedings*, 2020, vol. 2306, no. 1, p. 020035.
45. I. Prasojo, A. Maseleno, and N. Shahu, "Design of automatic watering system based on Arduino," *Journal of Robotics and Control (JRC)*, vol. 1, no. 2, pp. 59–63, 2020.
46. M. Asha T and M. Srija V, "Design and Implementation of Wireless Based Water Level Monitoring System using Arduino and Bluetooth," *Int. Resear. Jour. Engg. Tech*, vol. 7, 2020.
47. A. Kurniawan, "Arduino Nano 33 IoT Networking," in *Beginning Arduino Nano 33 IoT*, Springer, 2021, pp. 103–129.

Simulation of Groundwater Quality: Case Study of the Limestone Chain of the Western Rif of Morocco

Ghizlane Fattah, Fouzia Ghrissi, Jamal Mabrouki, and Saloua Elfanssi

1 Introduction

Since 1913, the water sector has oscillated between the supervision of public works, in charge of the public hydraulic domain, and that of agriculture, in charge of the irrigation policy. For many centuries, the populations of Morocco have lived in perfect symbiosis with the water environment, knowing how to take advantage of wet years and adapting to drought years by mitigating the impacts, thanks to an informed sense of economy in the use of water. The demand for water will increase by 50% by 2030, mainly due to the needs of industry, energy, and population growth. With population growth and economic development, the gap between water supply and demand in Morocco is widening [1].

Morocco is making significant efforts to mobilize water resources to cope with population growth and to support its economic and social development. However, due to urbanization, growing demographic pressure, insufficient wastewater treatment, deforestation in the country's main river basins, water resources used, or intended for drinking water production are increasingly exposed to multiple sources of pollution of urban, industrial, agricultural, and accidental origin [2]. The method

G. Fattah (✉) · F. Ghrissi
Civil Hydraulic and Environmental Engineering Laboratory, Water Treatment and Reuse Structure, Mohammadia School of Engineers, Mohammed V University in Rabat, Rabat, Morocco
e-mail: ghizlane_fattah@um5.ac.ma

J. Mabrouki
Laboratory of Materials, Nanotechnology and Environment, Department of Chemistry, Faculty of Sciences, Mohammed V University, Rabat, Morocco

S. Elfanssi
National Center for Research and Studies on Water and Energy (CNEREE), Cadi Ayyad University, Marrakech, Morocco

M. Azrour et al. (eds.), *IoT and Smart Devices for Sustainable Environment*, EAI/Springer Innovations in Communication and Computing,
https://doi.org/10.1007/978-3-030-90083-0_9

discussed here, which integrates water quantity and quality into water resource allocation, has the potential to provide important support to decision-makers facing these challenges [3].

The use of computers has given us the opportunity to better understand and evaluate our water resources through numerical model simulations and the verification of many plans and options. Numerical models, which are a combination of equations representing certain physical structures and the hydraulic impact they have on lakes and river flows, allow a user to evaluate the hydraulic conditions in the river basin and thus obtain a better understanding of human impacts on a natural or modified lake and river system. Research addresses this issue by including water quality in the allocation process, using programming models (simulations) to achieve optimal use of available water with different quality states [4, 5]. Models are calibrated by comparing calculated results to observed values and then making subsequent adjustments to the mathematical coefficients to obtain a better fit of the results [6]. Models are verified by comparing calculated results with historical data that are not used in the calibration. Simulation can replace experimental methods, which are costly and time-consuming [7].

Simulation models calculate the flows, levels, water quality, and other parameters based on physical laws and conceptual relationships. The simulation workflow requires many tasks that must be completed before programming can begin. Completion of these tasks defines the complexity of the model required, the data needed, the development time, the complex problems to be solved, and the expected model run time. In fact, these tasks are fundamental to building a simulation and implicitly guide the choice of model to be used [8]. Selecting a model that is too simple can lead to a lack of accuracy and certainty in decision making, while selecting a complex model can lead to misallocation of available resources, delays in studies, and an increase in required metrics and costs [9].

Recently, many water quality models have been widely applied worldwide, such as QUAL2E, QUAL2K, IberWQ, SWAT, and Vensim PLE [10, 11]. For the QUAL2K, which is the modern version of the QUAL2E system recommended by the US Environmental Protection Agency (EPA), is used to evaluate the self-cleaning capacity of US rivers that receive wastewater from municipal sources [12]. Quality2E is designed to analyze the effects of permanent pollution sources in accordance with American standards [13].

The modeling of the sources of unstable pollutants, which vary from one section of a river to another, is complicated by using the QUAL2E model, the QUAL2E module concerning the kinetics of the physicochemical balances and the transformations of the nutrients within the river, and, finally, the integration of a GIS interface [14, 15]. The majority of available water quality models have another downside, namely that their internal structure cannot be modified by users. The model includes empirical equations and coefficients that capture the specific conditions for which the model was first developed. As a result, users are not allowed to introduce equations or drivers that better describe a particular case studied [16]. The Eutrophication module of the QUAL2K model has been entirely embedded in the most recent versions of WASP, the model that is administered and maintained by the USEPA, to allocate different kinetic coefficients for different sections along a river is not permitted. This becomes a serious problem when effluents with different levels of biodegradation are released at different points in the river. Assuming a single biodegradation coefficient for water flow modeling in a river may underestimate the actual impact, reduce its power, and increase the margin of error in the modeling [17]. In the face of these limits of the already existing computer models, the design of a simulation model that permits the manipulation of the base equations of the water treatment model is needed. The model we tested on the river of the Rif chain, located in southern Morocco. This river is an appropriate case for study since its water quality has recently been degraded and needs to be tested more frequently, while our ability to take the amount of sampling action along the river is restricted due to the rough terrain and limited financing. Selection of a complex model may result in misdirection of the available resources, study time delays, and higher costs for required parameters [18].

2 Study Area

The study area is located in the heart of the Tingitana Peninsula, more precisely north of the city of Tetouan on the western Rif, which is made up of a mountain range stretching out over the Mediterranean Sea. They are characterized by the presence of rivers constituting small watersheds.

Geologically, it is formed as a tectonic unit made up of the fragmented and nested structural units of limestone and dolomite from the Triassic and Early Jurassic [19].

This limestone chain is characterized by a humid climate and by an intense karstification system that has generated several karsts ranging from limestone causeways to pits, caves, dry valleys, and travertine deposits (constitutes aquifers which benefit from the infiltration of rainwater [20].

3 Materials and Methods

The majority of physical, chemical, or biological phenomenon derived from natural environments can be characterized by complex systems of partial differential equations. The mathematical study of these equations is not an easy task and sometimes even impossible [21].

In this work, we focused on the numerical approximation of a mass balance equation using the finite difference (Eq. 1). With this approximation, we sought to simulate the nitrogen, phosphate, and oxygen cycles of the process. This approximation transforms a continuous problem governed by differential equations into a discrete one; so using an initial concentration of C0 (which should be known), we can deduce other concentrations at different times and locations

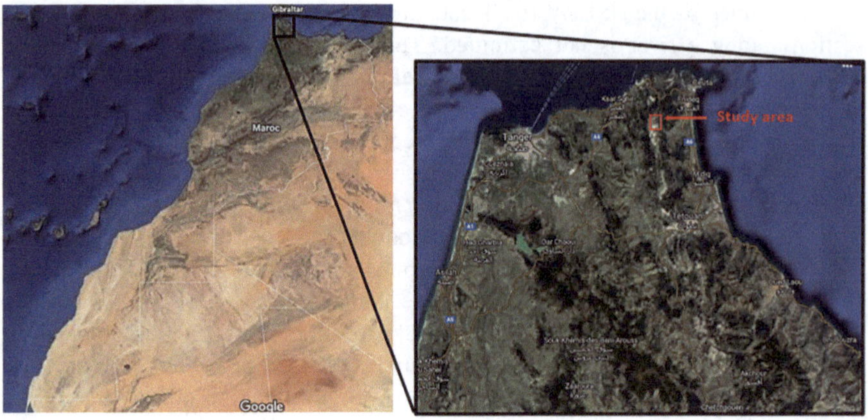

Fig. 1 Study area location

$$\frac{dC}{dt} = D_X \frac{d^2C}{d_X 2} - V_x \frac{dC}{dx} - KC \pm S_{\text{Interne}} \quad (1)$$

With

K is the reaction coefficient of the speed
D_x is the longitudinal dispersion Coefficient
S_{interne} is source terms
V_x is the flow velocity in the spatial direction x.

The individual response coefficients vary over defined ranges, enabling our model to be tuned to fit the conditions and the characteristics of the study area. The source terms (gain or loss) are associated with the phenomenon that influences the concentration of the parameters. These sources are important when liquid loads are dumped into the stream (domestic, agricultural, and industrial fluids) [22].

The calibrated model was tested only in the downstream region that has suffered a recent severe degradation. Different solutions were tested, allowing the study of possible enhancements to reduce the water quality degradation. The geometry of the configurations consisted of seven sections (Fig. 1), with varying dimensions. Each section was named after the nearest village or locality.

3.1 General Water Quality Status

To estimate the overall water quality in the lower portion, the water quality index was derived using simulated data and laboratory readings. This index ranks water quality on a 0–60 scale, where the 20–50 range represents very poor water quality status, while 0–25% represents excellent water quality. The NSF-WQI was calculated using the arithmetic index method [23] using the equation:

$$\mathrm{NSF-WQI} = \sum_{n=1}^{n} q_n W_n / \sum_{n=1}^{n} W_n \quad (2)$$

Where q_n (Eq. 2) is the grade that mirrors the relative value of the variables with the permitted standard value [24]. The $_{qn}$ is derived from the ideal value (V_i), which is equal to zero $V_i = 0$, except for pH ($V_i = 7$), temperature (T) ($V_i = 25$) and DO ($V_i = 14.6$ mg/L), the observed value (V_n), and the standard of the measured parameters allowed value (S_n).

$$q_n = 100\ [V_n - V_i] / [S_n - V_i] \quad (3)$$

Wn is the unit burden of the nth water quality characteristic [25], it is calculated using Eq. 4:

$$W_n = k / S_n \quad (4)$$

Where K is the proportionality constant and is determined by the formula 5:

$$K = 1 / \left(\sum 1\,(S_n = 1, 2, \ldots, n) \right) \quad (5)$$

Since the temperature (T), pH, electrical conductivity (EC), and water hardness (hydrotimetric titer TH) are not governed by the mass balance equation, we used stepwise regression methodology to reveal the existing simulated parameter correlation. This Regression method estimates the necessary parameters for the WQI-FNS using a combination of a simulator.

4 Results and Discussion

4.1 *Coefficients of Biochemistry Action*

The monitor was then calibrated using different biochemical reaction coefficients. These coefficient ranges are well known to the models of water quality [26]. The variation in these coefficients made the model results similar to the results obtained by the lab analysis. The values of the coefficients were checked on the basis of measured data (Table 1).

Table 1 Simulated sections geometry characteristics

Puits	*X*	*Y*	Piezometry (m)
P1	495 626	576 766	5.5
P2	495 361	576 856	6
P3	496 105	576 376	4.5

Table 2 Characteristics of water at the sampling points

Parameters	P1	P2	P3
T (°C)	16.3	14.2	16.2
pH	7.48	7.84	6.37
EC (μs/cm)	622	1750	360
O_2 (mg/L)	8.51	9.8	8.54
Salinity (g/L)	0.2	0.8	0.1
BOD_5 (mg/L)	122	147	160
Cl^- (mg/L)	1023.9	8423.6	1930.4
Turb (NTU)	2.27	5.68	4.89
HG (mg/L)	–	0.85	–
SS (mg/L)	2	4	2
P (mg/L)	17.1	19.9	16.9
NH_4^+ (mg/L)	1.2	1.4	1.1
NO^{3-} (mg/L)	0.2	0.4	0.3
NO^{2-} (mg/L)	5.6	5.8	5.7
PO_4^{-3} (mg/L)	0.4	0.8	0.7

4.2 Simulation and Monitoring of Concentrations

The performance of our model is reported for the rainy and dry months. As shown in the figure, there was a strong relationship between the simulated and measured data, except for P1 in which the model underpredicted the field measured parameters. This discrepancy could be attributed to the dilution regressed when the rivers, or that the location where we chose to characterize the river was not well chosen. Together, our model suggests that the two rivers are well mixed at P1, when in reality they are not (Table 2).

The biochemical oxygen demand (BOD) phosphate results were not well adapted to the field, while the rest of the simulated variables overlapped well with the measurements from the measurement laboratory. The results of the dry season validation are presented in Figs. 3 and 4. From these results, there was still a substantial difference between the simulated and measured results at the last sampling station. This may confirm that this station was chosen incorrectly and the two rivers are not yet well mixed at this location. The phosphate scores were less well-calibrated with the data measured; however, the BOD scores well compared to the rainy season. In general, the simulated BOD and phosphate results were acceptable (Table 3).

Table 3 Selected reaction coefficients

Parameter coefficient	Admitted value
K1: Bio-oxidation coefficient for biochemical oxygen demand (1/day)	0.64
K2: Aeration coefficient (1/day)	4.10
K3: Sedimentation effect coefficient (1/day)	0.6
Os: Oxygen saturation point	0.8
KA: Dissolved oxygen from chlorophyll A (1/day)	0.02
A1: Chlorophyll A phytoplankton (mg /L)	0.001
A2: Sessile chlorophyll A algae (mg /L)	0.001
Kkg: Organic nitrogen decomposition rate (1/day) 0.4	0.4
KNH4: Ammonium decomposition rate (1/day) 0.4	0.4
KNO2: Nitrate or nitrite decomposition rate (1/day)	0.1
KPO4_1: Orthophosphate decomposition rate (1/day)	0.2
KPO4_2: Orthophosphate consumed by algae (1/day)	0.1
KB: Nitrate consumed by algae (1/day)	0.1

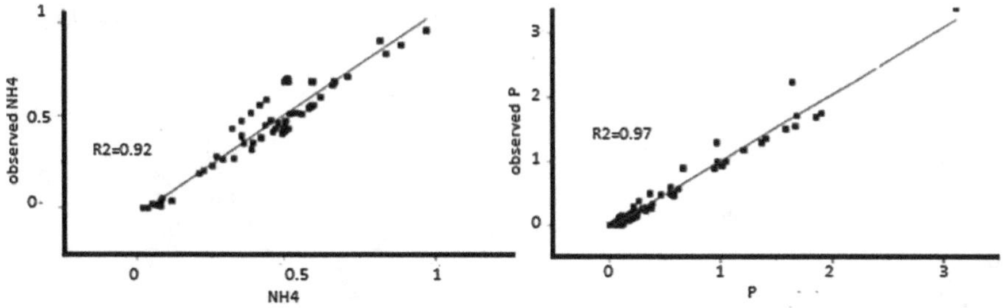

Fig. 2 Simulated and measured NH_{4+}, P correlation

4.3 *Stability Analysis*

The stability analysis ensures that the impacts of monthly variation on the suggested model are being minimized. This analysis is conducted by comparing the linear correlations between the measured and simulated EU parameters. The coefficient of definition (R^2) and root mean square error (RMSE) evaluate the efficiency of our simulation model and to compute the amount of error that could be created between observed data (Fig. 2).

The consistency test for simulated and measured parameters indicated a high correlation ($R^2 > 0.8$) for nitrogen and phosphate forms (Fig. 2). However, the results for dissolved oxygen (DO) and BOD were not as stable ($R^2 > 0.8$). This unsteadiness could be attributed to the pollution source loading from neighboring areas (Fig. 3).

This recharge was not well approximated due to the fact that there is no sewerage system at these sites. As shown in Table 4, the electrical conductivity can be computed from the calculated simulated parameters NH_4^+, NO^{3-}, and DO, while

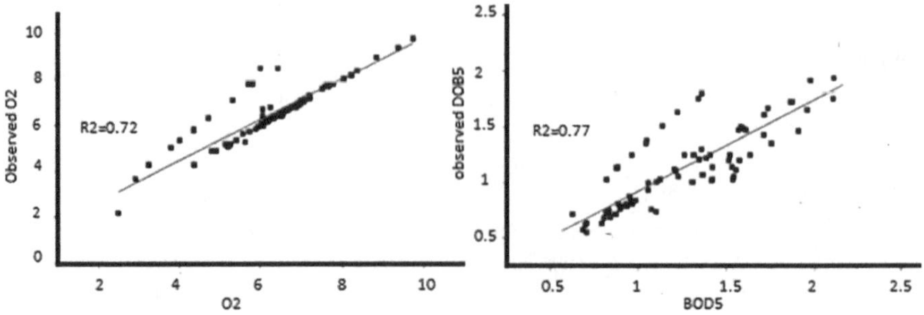

Fig. 3 Simulated and measured O2, BOD5 correlation

Table 4 Correlation between simulated parameters

	Correlated parameters	R^2	RMS
Electrical conductivity	NH_4^+, NO^{3-} and OD	0.62	29.3
Water hardness (TH)	NO^{2-} and PO_4^{3-}	0.64	7.02
pH	BOD, NO^{3-} and TH (estimated by NO^{2-} and PO_4^{3-})	0.75	0.16

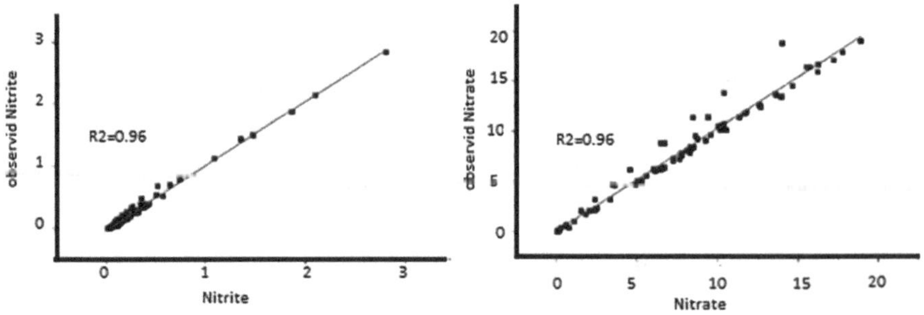

Fig. 4 Corrélation entre NO^{2-}, NO^{3-} simulés et mesurés

the pH can be calculated using a three-parameter combination (BOD, NO^{3-}, and water hardness) with high accuracy ($R^2 = 0.75$). Using the results of the correlation, we created an equation for estimating these parameters by using the simulated parameters.

On the basis of the WQI-NSF calculated using the parameters simulated, many water quality methods have been evaluated, where the quality average rate has been set at 50% in the WQI-NSF index. This evaluation process is valuable for testing the reliability, sustainability, and effectiveness of approaches to improve water quality impairment of water quality and is also less costly. The comparison between NSF WQIs that were derived from simulated and measured parameters over the entire data period was gathered revealed a high degree of correlation ($R^2 = 0.82$) (Fig. 5). This correlation showed the reliability of the simulated data for the NSF WQI computation was high (Fig. 4).

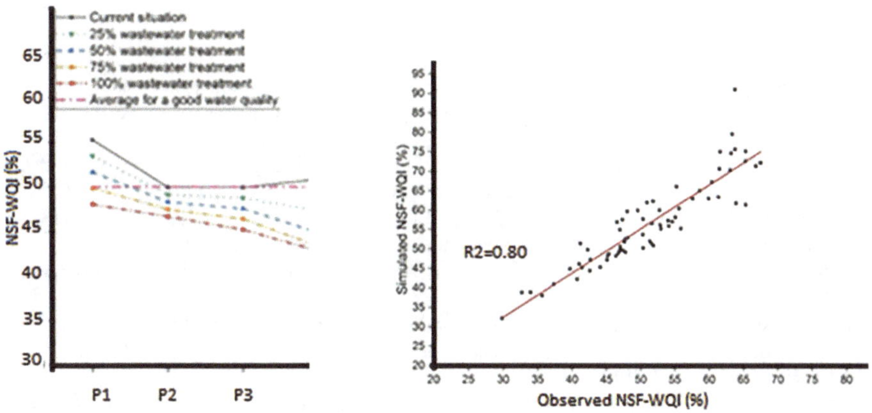

Fig. 5 Observed efficiency on NSF-WQI evolution

As previously mentioned, wastewater is dumped into the El Abid River, which has major risks in terms of quality deterioration. Therefore, the wastewater purification plant proposals, among others, were tested to assess their feasibility. Four scenarios were tested on the basis of wastewater treatment efficiency (30%, 45%, 80%, and 90%, which represents the complete treatment of the liquid discharge). The appropriate solutions were evaluated in terms of success in improving water quality and reducing costs. As shown in Fig. 5, it is not feasible to provide complete treatment of wastewater with 100% efficiency for the last two plants, P2, P3, which could be explained by other (natural) factors involved in the poor water quality (such as geological structures that are formed by evaporite rocks increasing the anion contained in the water) [27, 28]. In terms of effectiveness, the scenarios tested showed that treatment of liquid waste at the 25% level was insufficient to improve water quality conditions in all sections of the stream. However, when the level of treatment exceeded 50%, it significantly decreased the effects of the liquid discharges.

5 Conclusions

This process of simulation can reduce the analytical cost of individual samples and assist in the selection, planning, or optimization of these solutions by the decision-makers. Simulation by this chemical and the biological processes by solving mass balance equations is an effective tool that can help predict the quality of water and select appropriate and relevant solutions to the challenge of improving water quality. Testing our simulation model on areas suffering from degraded water quality showed that most of the areas are not affected by degraded water quality. The sustainable protection method consists of creating at least three wastewater treatment plants by natural lagooning with at least 50% efficiency for the P2-P3, while 75% is the minimum treatment level required at the site.

References

1. Hellegers, P., Immerzeel, W., & Droogers, P. (2013). Economic concepts to address future water supply–demand imbalances in Iran, Morocco and Saudi Arabia. Journal of hydrology, 502, 62–67.
2. Sowers, J., Vengosh, A., & Weinthal, E. (2011). Climate change, water resources, and the politics of adaptation in the Middle East and North Africa. Climatic Change, 104(3), 599–627.
3. Mabrouki, J., Benbouzid, M., Dhiba, D., & El Hajjaji, S. (2020). Simulation of wastewater treatment processes with Bioreactor Membrane Reactor (MBR) treatment versus conventional the adsorbent layer-based filtration system (LAFS). International Journal of Environmental Analytical Chemistry, 1–11.
4. Jayakrishnan, R. S. R. S., Srinivasan, R., Santhi, C., & Arnold, J. G. (2005). Advances in the application of the SWAT model for water resources management. Hydrological Processes: An International Journal, 19(3), 749–762.
5. Mabrouki, J., El Yadini, A., Bencheikh, I., Azoulay, K., Moufti, A., & El Hajjaji, S. (2018, July). Hydrogeological and hydrochemical study of underground waters of the tablecloth in the vicinity of the controlled city dump mohammedia (Morocco). In International Conference on Advanced Intelligent Systems for Sustainable Development (pp. 22–33). Springer, Cham.
6. Naoum-Sawaya, J., Ghaddar, B., Arandia, E., & Eck, B. (2015). Simulation-optimization approaches for water pump scheduling and pipe replacement problems. European Journal of Operational Research, 246(1), 293–306.
7. Knight, L. A., Pal, S., Coleman, J. C., Bronson, F., Haider, H., Levine, D. L., ... & Rullkoetter, P. J. (2007). Comparison of long-term numerical and experimental total knee replacement wear during simulated gait loading. Journal of biomechanics, 40(7), 1550–1558.
8. Rauch, S. L., Phillips, K. A., Segal, E., Makris, N., Shin, L. M., Whalen, P. J., ... & Kennedy, D. N. (2003). A preliminary morphometric magnetic resonance imaging study of regional brain volumes in body dysmorphic disorder. Psychiatry Research: Neuroimaging, 122(1), 13–19.
9. Chapra, M. U., Khan, S., & Al Shaikh-Ali, A. (2008). The Islamic vision of development in the light of maqasid al-Shariah (Vol. 15). Iiit.
10. Debele, B., Srinivasan, R., & Parlange, J. Y. (2008). Coupling upland watershed and downstream waterbody hydrodynamic and water quality models (SWAT and CE-QUAL-W2) for better water resources management in complex river basins. Environmental Modeling & Assessment, 13(1), 135–153.
11. Gao, L., & Li, D. (2015). A review of hydrological/water-quality models. Frontiers of Agricultural Science and Engineering, 1(4), 267–276.
12. Mabrouki, J., Moufti, A., Bencheikh, I., Azoulay, K., El Hamdouni, Y., & El Hajjaji, S. (2019, July). Optimization of the Coagulant Flocculation Process for Treatment of Leachate of the Controlled Discharge of the City Mohammedia (Morocco). In International Conference on Advanced Intelligent Systems for Sustainable Development (pp. 200–212). Springer, Cham.
13. Brown, L. C., & Barnwell, T. O. (1987). The enhanced stream water quality models QUAL2E and QUAL2E-UNCAS: documentation and user manual (Vol. 600, pp. 3–87). EPA.
14. Birgand, F., Lefrancois, T., Grimaldi, C., Novince, E., Gilliet, N., & Odoux, C. G. (2004). Mesure des flux et échantillonnage des matières en suspension sur de petits cours deau. Ingénieries eau-agriculture-territoires, (40), p. 21.
15. Srinivasan, R., & Arnold, J. G. (1994). INTEGRATION OF A BASIN-SCALE WATER QUALITY MODEL WITH GIS 1. JAWRA Journal of the American Water Resources Association, 30(3), 453–462.
16. Li, X., Gao, X., Shi, W., & Ma, H. (2014). Design strategies for water-soluble small molecular chromogenic and fluorogenic probes. Chemical reviews, 114(1), 590–659.
17. de SI Gonçalves, J. C., & Giorgetti, M. F. (2013). Mathematical model for the simulation of water quality in rivers using the VENSIM PLE® software. Journal of Urban and Environmental Engineering, 7(1), 48–63.).
18. Yu, L., & Salvador, N. N. B. (2005). Modeling water quality in rivers.).

19. Chapra, M. U. (2008, October). The global financial crisis: can Islamic finance help minimize the severity and frequency of such a crisis in the future. In A paper presented at the Forum on the Global Financial Crisis at the Islamic Development Bank on (Vol. 25).
20. El Gharbaoui, A., 1981. La terres et l'homme dans la péninsule tingitane/étude sur l'homme et le milieu naturel dans le Rif Occidental. Trav. Inst. Sci. serie Geol. Geogr. Phys. 15, 1–439 (in french).
21. Plan directeur intégré des ressources en eau des bassins du Loukkos, du tangérois et côtiers méditerranéens, 2012, p 31
22. Shampine, L. F. (2003). Singular boundary value problems for ODEs. Applied Mathematics and Computation, 138(1), 99–112.
23. Mirbagheri, S., Abaspour, M., & Zamani, K. (2009). Mathematical modeling of water quality in river systems. European Water, 27(28), 31–41.
24. Smith, D. E. (1990). The conceptual practices of power: A feminist sociology of knowledge. University of Toronto Press.
25. Rao, A. V., Benson, D. A., Darby, C., Patterson, M. A., Francolin, C., Sanders, I., & Huntington, G. T. (2010). Algorithm 902: Gpops, a matlab software for solving multiple-phase optimal control problems using the gauss pseudospectral method. ACM Transactions on Mathematical Software (TOMS), 37(2), 1–39.
26. Sapkal, R. S., & Valunjkar, S. S. (2013). TECHNOLOGY (IJCIET). Journal Impact Factor, 4(4), 119–134.
27. Chapra 2008; de SI Gonçalves et F Giorgetti 2013; Rauch et al. 1998; Yu et Salvador 2005.
28. Fattah, G., Ghrissi, F., Mabrouki, J., & Kabriti, M. (2021). Control of physicochemical parameters of spring waters near quarries exploiting limestone rock. In E3S Web of Conferences (Vol. 234, p. 00018). EDP Sciences.

A Novel Anomaly Network Intrusion Detection System for Internet of Things Security

Azidine Guezzaz, Said Benkirane, and Mourade Azrour

1 Introduction

Recently, The Internet of Things (IoT) has known an important progress and development. Hence, IoT is transforming many daily services [1, 2, 3] such as health care, environment, telecommunications, and many others. The IoT can be denoted as a large network interconnecting various objects in real or virtual words, computer systems, and people. These devices are able to collect, analyze, and communicate data from one place to another.

Nowadays, the main concern of researchers is to improve the security issues to design and implement strong approaches to face threats and vulnerabilities around Internet of Things environment. Hence, the computer science security allows a set of hardware tools and software, like intrusion detection systems and authentication [4], as an efficient and enhanced approaches aiming to secure and protect traffic circulated within systems and networks.

Intrusion detection is one of spread and useful technologies developed to stop unauthorized uses of systems. It is a recent alternative approach to defend various resources and empower computer security. IoT security is complicated because of system complexity and device heterogeneity. The main goal of this work is to model and validate a new hybrid approach of anomaly detection using Naïve Bayes algorithm to monitor and analyze to more secure data events of Internet of Things environment.

A. Guezzaz (✉) · S. Benkirane
Technology Higher School, Cadi Ayyad University, Marrakesh, Morocco

M. Azrour
Department of Computer Science, Faculty of Sciences & Techniques, Moulay Ismail University, Errachidia, Morocco

M. Azrour et al. (eds.), *IoT and Smart Devices for Sustainable Environment*,
EAI/Springer Innovations in Communication and Computing,
https://doi.org/10.1007/978-3-030-90083-0_10

The reminder of the paper is organized as follows. In the second section, we present a related works proposed in the literature. The third section gives background information about Internet of Things and intrusion detection technologies and presents also certain machine learning techniques, such as Naïve Bayes. The new approach is designed in the fourth section; it details various components of our proposed model. The experimental study and results of our model are discussed in the fifth section. Finally, the paper is achieved with a conclusion and future works.

2 Related Work

The usefulness of networks and computer systems in life has grown considerably. Their usage is becoming an interesting and necessary task in society. In addition, many organizations look for developing the technology by producing several software security exploited in various fields: military, health, and electronic commerce.

The vulnerabilities represent different weaknesses in software or hardware that can be exploited by intruders and attacks to access a network [9, 10]. Furthermore, they signify an absence of security procedure that could be exploited by a threat. Therefore, they can infect and destruct all systems.

So, to secure like environment, the explored vulnerabilities in various layers are taught into account to find many solution security for each layer [5, 6]. For this reason, the IoT environment has to do real-time management to secure resources and data. This real-time management requires access to real-time data from all network sensors. Among the tools that are implemented for security threats and monitoring, we find intrusion detection systems [7]. According to the complexity and the increasing of attacks on networks, the information represents a precious thing, which must be protected.

Accordingly, computer security is an essential issue, which ensures the hardware or software resources of an organization to be used only within an authorized framework by aiming at the following main objectives: authentication, integrity, confidentiality, availability [10]. Intrusion detection systems ensure reliable threat detection in order to secure systems. They are widely studied in industry and academia. They are used to detect intruders aiming to cause danger to computer systems, alert the security administrator of any suspicious or illegal activity and often make decisions to anticipate attacks [11, 12, 13, 19]. The majority of intrusion detection systems suffer from a large generation of false positive and some false negative, which infects their effectiveness and accuracy. The IoT environment provides a set of services to organization and individuals using an Internet connection to access these services. Hence, the automatic detection of intrusions within these environments becomes more useful and necessary [23, 24]. A set of search work present try to propose and validate a sophisticated intrusion detection system to enhance the security level of networks especially those of Internet of Things [6, 7].

3 Background

3.1 Internet of Things Security

The Internet of Things refers to enormous system that interconnects innumerable things both in physical and virtual words. Internet of Things networks are characterized by three-layer classification consisting of application, perception, and network layers. The architecture of an Internet of Things system consists of several layers communicating with each other to connect the real world of objects to the virtual world. However, there is no global architecture agreement. Therefore, the various projects do not always follow the same formal architecture. For this reason, researchers have proposed much several architectures. According to some research works, the architecture of the Internet of Things consists of three layers, but others support four layer architecture. Due to the improvement of the Internet of Things, the three-layer architecture cannot meet the requirements of applications. The five layer architecture (Fig. 1) has also been proposed because of a challenge posed by the Internet of Things.

3.2 Machine Learning Techniques

A machine learning technique is a subfield of artificial intelligence that aiming to improve design and use of different technologies [8, 14, 16, 17]. The classification is a part of machine learning tasks that tries to categorize instances of dataset provided to handle given problems. Any instance is assumed to belong to a fixed class. The class of an instance is determined by a class attribute. All instances are used to build model. The training phase aims to build a classifier model and uses several algorithms to solve classification tasks by combining classes of similar objects. This is a supervised learning, which imposes labeled data in the training. It contains also the regression tasks. In this case, the training is process in which the model changes until a desired one [10, 22].

- SVM Algorithm

Support vector machine (SVM) is supervised machine learning technique that used for classification task. It is implemented to handle not only discrimination problems but also that of regression. The main objective of SVM is to transform inputs into a characteristic dimensional space to discover a useful hyperplane, which separates two groups by representing data in space. The linear border produces a maximum margin. The discovered hyperplane represents a solution to a constrained optimization task. To search nonlinear separating surfaces, it is useful to introduce a kernel, which encodes nonlinear transformation of data. Hence, the equations are found using certain scalar products with kernel and some database weights [15, 21].

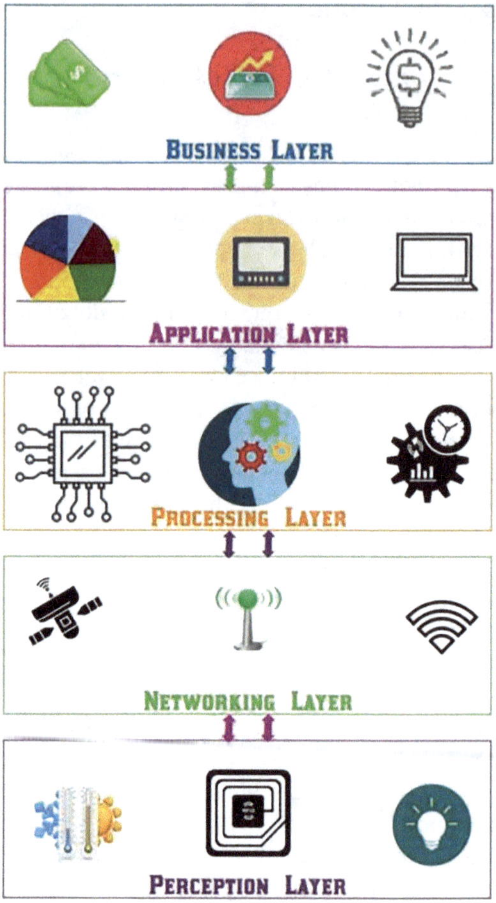

Fig. 1 Five layers of IoT architectures

- K-NN Algorithm

K-nearest neighbors (K-NN) is the most widely used algorithm for the classification task. Typically used by continuous attributes but can be changed and adapted for qualitative attributes. The goal is to predict classification of new instance, called also, unseen instances, based on one or more instances. In practice, if we have several in- stances, it is tedious to predict the class, but if the same principle will be applied. Make decisions by researching one or more similar cases that have already been resolved [10].

- NB Algorithm

Naïve Bayes (NB) is an algorithm that uses probability theory, essentially used for classification tasks. It tries to find the most probable of the possible classifications. The Bayesian method is expensive in execution time, but it provides optimal solutions. A classification problem under Naive Bayes can be formulated using the posterior probabilities [15, 21].

P (C | X) is a probability for an instance $X = (x_1, ..., x_k)$ belongs in class C. The goal is to assign an instance X to the class C such that P (C | X) is maximum.

$$P\ (C \mid X) = P\ (X \mid C)\ P\ (C)\ /P\ (X)$$

P (X) represents a constant for classes.

P (C) represents relative frequency of instances of class C such that P (C | X) is maximal.

C such that P (X | C) · P (C) is maximal.

The constraint is that the computation of P (X | C) is not feasible.

The Naïve Bayes prefers to deal with independence of attributes:

$$P\ (x_1, \ldots, x_k \mid C) = P\ (x_1 \mid C) \cdot \ldots \cdot P\ (x_k \mid C)$$

P (x_i | C) is predicted as relative frequency of instances with the value x_i in class C.

The data quality is considered as a useful field in data science that aiming to obtain high quality of data by using a set of operations and techniques to facilitate decision making. The preprocessing tasks allow much of algorithms that are used to increase the training accuracy and high quality of data [18, 21, 22]. Thus, before to choose an algorithm, it is useful to understand dataset well. To choose a training algorithm, we take into account the specification of problem to be solved. In fact, the classification task becomes a very interesting field that tries to solve complex problems. The data activities are not always ready for a good analysis. Thus, the standardization of unstructured data requires an enhanced work.

Recently, traditional techniques are not capable to provide an accurate IoT security. Our work is focused on intrusion detection and prevention system to improve security level of IoT environment. Its main goal is to design and validate a new intrusion detection approach using machine learning techniques in order to improve the data quality and detection rate to better secure an IoT environment.

4 New Approach

This section describes various elements of proposed approach, deployment steps, and several techniques that we implement to make a rigorous framework. The design of our approach is described below (Fig. 2):

Our model includes a number of components and stages that start with collection of traffic from various sensors that monitor IoT environment. Enough information about environment are gathered after. The next step is setting up to analyze collected events:

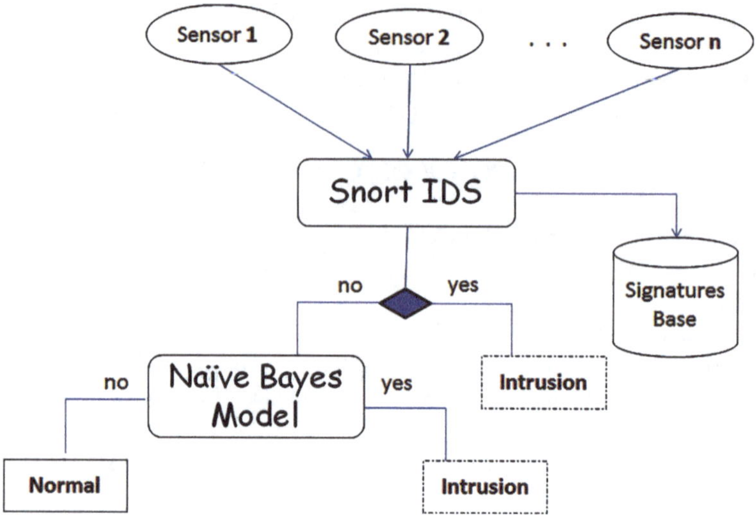

Fig. 2 Proposed hybrid approach based on Snort and Naïve Bayes model

This design is based on hybrid detection approach, which uses Snort intrusion detection to satisfy misuse detection. The gathered activities within various sensors are compared by contents of signatures base to control the first detection. The anomaly detection is used to improve detection rate of proposed approach.

Data collection process is satisfied by a set of sensors as computer software that used to intercept traffic from various devices; for this, we utilize the Snort intrusion detection system, not only to collect traffic but also to enhance the detection rate of our approach. We use Snort to handle first intrusion detection based on misuse detection. The second detection will be handled by our proposed classifier based on Naïve Bayes algorithm.

5 Experimental Study and Assessments

In this section, we present an experimental work of our contribution for intrusion detection approach to improve an IoT environment security. We describe in details the implementation steeps and suggested solutions to validate the approach. The dataset assessment plays a main role of the validation step of any intrusion detection approach. It allows us to evaluate usefulness and capability of various proposed method in intrusion detection [20]. Due to privacy issues, the datasets that are used for network packet analysis in commercial products are not available. However, there are some public datasets such as DARPA, KDD, and NSL-KDD that are used widely to train and build models. The earliest effort to create an IDS dataset was made by DARPA (Defense Advanced Research Project Agency) in 1998, and they created the KDD 98 (Knowledge Discovery and Data Mining (KDD)) dataset. The KDD cup 99 datasets consist of training and test set. There are 4,940,000 data sample in the training set; these samples are distributed between normal behavior and 24 attacks. There are 311,029 data samples that comprise normal network traffic and 38 types of attack; 24 attacks existed in training set in addition to 14 new attacks. Those attacks fall into four main categories:

- DOS: denial-of-service
- R2L: unauthorized access from a remote machine
- U2R: unauthorized access to local superuser (root) privileges
- Probing: surveillance and other probing

It is also important to mention that the KDD training data set contains a large number of connections for the categories normal, probe, and DoS. They represent approximately 99.76% of the whole dataset.

The empirical study is implemented on a computer Intel (R) Core (TM) i5-2520M CPU @2.50 GHz RAM 8,00 Go Windows 7 professional 64 bits. The model training is implemented using python version (Python 3.8.0). To build and test our model, we use a part data set from KDDCUP 99.

There are four possible exits:

- True negative (TN), which presents the number of normal instances predicted as normal
- True positive (TP), which presents the number of attacks that are attack
- False positive (FP), which presents the number of attacks predicted as normal
- False negative (FN), which presents the number of normal instances predicted as attack

These performance evaluations are not dependent on the size of the training and testing sets and can be really helpful in assessing the performance of the ensemble model:

- Accuracy (ACC) calculates how the classifier is correct:

$$\mathrm{ACC} = \frac{\mathrm{TP} + \mathrm{TN}}{\mathrm{TP} + \mathrm{TN} + \mathrm{FP} + \mathrm{FN}}$$

- Detection rate (DR) is a performance metric, which indicates the ratio of the number of samples that are correctly classified as attack to the total number of attack samples:

$$\mathrm{DR} = \frac{\mathrm{TP}}{\mathrm{TP} + \mathrm{FN}}$$

- Precision calculates the percentage of intrusions that have occurred, and the IDS detect them correctly:

Table 1 Metric performances of SVM, KNN, and proposed model

	SVM	K-NN	Proposed model
ACC	93%	94.4%	96.45%
DR	91.25%	92%	93%
Precision	93.5%	93%	94%
F-score	92.36%	92.49%	93.49%

$$\text{Precision} = \frac{\text{TP}}{\text{TP} + \text{FP}}$$

- F-Score is a statistical criterion to evaluate the performance of the systems. In fact, this criterion is a harmonic mean between metrics recall and precision:

$$\text{F–score} = \frac{2 \times \text{DR} \times \text{Precision}}{\text{DR} \times \text{Precision}}$$

There is no standard detection system able to overcome all vulnerabilities of actual systems. The down researches in intrusion detection try to minimize in maximum those vulnerabilities by proposing optimal architectures integrating a sophistical techniques and algorithms.

To clarify, the obtained results are described and summarized in Table 1. The detection rate of our proposed approach exceeds 93 %, while that of SVM and KNN achieve 91.25% and 92%, respectively. In terms of precision, the new model realizes 94%, 93.5 %, for SVM and 93% for KNN.

The comparison results in Table 1 confirms that the new proposed model presents various advantages and accurate results according to other methods SVM and KNN, in terms of accuracy, precision, and detection rate.

6 Conclusion and Future Works

Intrusion detection approaches are reliable technologies used to empower security issues by monitoring traffic handled by systems within networks. In this work, we design and validate a hybrid intrusion detection approach based on Snort IDS and Naïve Bayes classifier to control and classify IoT network traffic. The new approach is an anomaly detection that implements a binary Naïve Bayes classifier to distinguish between normal data and intrusion. A data quality process is held before classification using Naïve Bayes techniques because of data sources heterogeneity to increase accuracy rate of our model. The present contribution improves detection rate of designed model and provides good performances and advantages. The future work aims to integrate some efficient techniques of features engineering to choose and select accurate attributes presented to enhance our model.

References

1. J. Mabrouki, M. Azrour, and S. El, "Use of Internet of Things for monitoring and evaluation water's quality: Comparative study," International Journal of Cloud Computing, 2021.
2. J. Mabrouki, M. Azrour, Y. Farhaoui, and S. El Hajjaji, "Intelligent system for monitoring and detecting water quality," in Big data and networks technologies, vol. 81, Y. Farhaoui, Ed. Cham: Springer International Publishing, 2020, pp. 172–182.
3. H. Kim, H. Choi, H. Kang, J. An, S. Yeom, and T. Hong, "A systematic review of the smart energy conservation system: From smart homes to sustainable smart cities," Renewable and Sustainable Energy Reviews, vol. 140, p. 110755, Apr. 2021, doi: https://doi.org/10.1016/j.rser.2021.110755.
4. M. Azrour, J. Mabrouki, A. Guezzaz, and Y. Farhaoui, "New enhanced authentication protocol for Internet of Things," Big Data Mining and Analytics., 4, 1, pp. 1–9, Mar. 2021, doi: https://doi.org/10.26599/BDMA.2020.9020010.
5. S. Nasiri, F. Sadoughi, M. HTadayon, and A. Dehnad, Security requirements of Internet of Things-based healthcare system: A survey study. Acta Informatica Medica 2019 Dec; 27(4): 253–258. doi: https://doi.org/10.5455/aim.2019.27.253-258.
6. M. U. Farooq, et al, A review on Internet of Things (IoT). International Journal of Computer Applications (09758887) 113 1, March 2015.
7. Ahmed B. Altamimi and Rabie A. Ramadan. Towards internet of things modeling: A gateway approach. Complex Adaptive System Modeling (2016) 4:25. DOI https://doi.org/10.1186/s40294-016-0038-3
8. A. Guezzaz, Y. Asimi, M. Azrour and A. Asimi. Mathematical validation of proposed machine learning classifier for heterogeneous traffic and anomaly detection", Big Data Mining and Analytics ISSN 2096-065403/07 pp 18–24 4, 1, DOI:https://doi.org/10.26599/BDMA.2020.9020019 2021.
9. Omar Achbarouet al, "Cloud security: A multi agent approach based intrusion detection system", Indian Journal of Science and Technology, 10(18), https://doi.org/10.17485/ijst/2017/v10i18/109044, 2017.
10. Guezzaz, A., Asimi, Z., Batou, Y., Asimi and Y. Sadqi. A global intrusion detection system using PcapSockS sniffer and multilayer perceptron classifier; International Journal of Network Security (IJNS), 21, 3, PP. 438-450, https://doi.org/10.6633/IJNS.201905_21(3).10). 2019.
11. China, R., Avadhani, P.: A comparison of two intrusion detection systems. IJCST 4, 1, ISSN: 0976-8491 (2013).
12. Santos, B., Chandra, T., Ratnakar, M., Baba, S., Sudhakar, N.: Intrusion detection system-Types and prevention. International Journal of Computer Science and Information Technologies, 4 (1) (2013).
13. A. Guezzaz, A. Asimi, Y. Asimi, Z. Tbatou and Y. Sadqi. A lightweight neural classifier for intrusion detection. General Letters in Mathematics, 2, 2, pp.57-66, (2017).
14. Guezzaz, A. Asimi, M. Azrour, Z. Batou and Y. Asimi. A multilayer perceptron classifier for monitoring network traffic, Big Data and Networks Technologies 2020.
15. A. Guezzaz, A. Asimi and Y. Asimi. A hybrid NIPS based on PcapSockS sniffer and neural MLP, advances in intelligent systems and computing. Springer. ISBN 978-3-319-64718-0 (2017).
16. Bychkov D, Linder N, Turkki R, Nordling S, Kovanen PE, Verrill C, et al. Deep learning based tissue analysis predicts outcome in colorectal cancer. Scientific Reports 2018;8:3395.
17. A. Buczak and E. Guven, A survey of data mining and machine learning methods for cyber security intrusion detection, IEEE Communication Surveys and Tutorials, 18, 2, 2016.
18. I. Cvitićl, D. Peraković, M. Periša and B. Gupta, Ensemble machine learning approach for classifcation of IoT devices in smart home, International Journal of Machine Learning and Cybernetics https://doi.org/10.1007/s13042-020-01241-0.
19. S.X.Wuand W.Banzhaf, "The use of computational intelligence in intrusion detection systems: A review," Applied Soft Computing., 10, 1,pp. 1–35, 2010.

20. A. Khraisat, I. Gondal, P. Vamplew and J. Kamruzzaman, Sur- vey of intrusion detection systems: techniques, datasets and challenges, Cybersecurity (2019), https://doi.org/10.1186/s42400-019-0038-7.
21. J. Gu, S. Lu, An effective intrusion detection approach using SVM with naïve Bayes feature embedding, 4048(20)30431-4, https://doi.org/10.1016/j.cose.2020.102158, Computers & Security (2020).
22. Anna L. Buczak and Erhan Guven, A survey of data mining and machine learning methods for cyber security intrusion detection, IEEE Communication Surveys and Tutorials, 18, 2, Second quarter 2016.
23. Ansam Khraisat, Iqbal Gondal, Peter Vamplew and Joarder Kamruzzaman, Survey of intrusion detection systems: Techniques, datasets and challenges, Cybersecurity (2019), https://doi.org/10.1186/s42400-019-0038-7.
24. Mohamed Amine Ferrag, Leandros Maglaras, Sotiris Moschoyiannis, Helge Janicke, Deep learning for cyber security intrusion detection: Approaches, datasets, and comparative study. Journal of Information Security and Applications 50 (2020) 102419.

Precision Agriculture: Assessing Water Status in Plants Using Unmanned Aerial Vehicle

Kaoutar Lkima, Francisco Pedrero Salcedo, Jamal Mabrouki, and Faissal Aziz

Introduction The term "precision irrigation" commonly refers to water according to plant necessity in the appropriate place, duration, and quantity. Utilizing variable rate of water application is possible with effective systems using current controllers in irrigation machines can be attainable in the market. Nevertheless, this application requires data from space and time inadequate quality and appropriate accuracy of water status in vegetation cover, achievable primarily through aerial monitoring and images [1, 2, 3]. The digitized aerial image turned to be inexpensive and publicized because of the availability in the market with a small price, severe results, and unmanned aerial vehicles of small size holding high-quality digital cameras, and

K. Lkima (✉)
Laboratory of Water, Biodiversity & Climate Change, Semlalia Faculty of Sciences, University Cadi Ayyad, Marrakech, Morocco

National Centre for Research and Studies on Water and Energy (CNEREE), University Cadi Ayyad, Marrakech, Morocco

Department of Irrigation, CEBAS-CSICCampus Universitario de Espinardo, Murcia, Spain

F. P. Salcedo
Department of Irrigation, CEBAS-CSICCampus Universitario de Espinardo, Murcia, Spain

J. Mabrouki
Laboratory of Materials, Nanotechnology and Environment, Department of Chemistry, Faculty of Sciences, Mohammed V University, Rabat, Morocco

F. Aziz
Laboratory of Water, Biodiversity & Climate Change, Semlalia Faculty of Sciences, University Cadi Ayyad, Marrakech, Morocco

National Centre for Research and Studies on Water and Energy (CNEREE), University Cadi Ayyad, Marrakech, Morocco
e-mail: f.aziz@uca.ma; faissalaziz@gmail.com

M. Azrour et al. (eds.), *IoT and Smart Devices for Sustainable Environment*,
EAI/Springer Innovations in Communication and Computing,
https://doi.org/10.1007/978-3-030-90083-0_11

examining these recent conductive technologies precision irrigation data acquiring is required.

Satellite multispectral imageries (MSI) attributable to sensor characteristics and high-altitude platforms (HAPs). These satellites are covering vast territories. Furthermore, several satellite programs such as Landsat, Modis, Aster, SPOT, Sentinel-1, and Sentinel-2 currently offer free data, therefore encouraging satellite images employed in effective agricultural applications [4, 5]. Furthermore, new techniques and algorithms employing satellite or unmanned aerial vehicle integrated multispectral imaging have been created for the vegetation of pixel classification [6, 7]. Unmanned aerial vehicles as low altitude platforms provide images with high spatial quality and flexible flight schedule [8], allow distinction between clear canopy pixel and elements in the environment [9, 10], and categorize various elements detailed in canopy [11, 12].

There have been many investigations at various phenological phases on the spectral reflectance of the plants in the visible and near-infrared spectrums as an element of electromagnetic radiation. It is possible to use data at the canopy level to have such a sufficient understanding of the spectral reaction of the leaf and the operation through this tier [13]. The use of spectral indicators helps improve data obtaining through remote sensing information because they improve intended obtained characteristics by diminishing the outcome soil, view angle, and topography by diminishing in outcome view angle, topography, and soil [14].

As an alternative way to identify water potential, canopy temperature measurement was proposed [15]. Basing on the water tension caused by closed stomata decreased transpiration and evapotranspiration cooling [16]. The indicator of water tension in plants, which is determined as crop cover water tension index CWSI was recommended to be identified [17]. The use of infrared sensors or imaging for detecting the canopy temperature has exhibited excellent ability for the calculation of CWSI also has shown exceptional performance for water status in the plant assessment in various crops has seen in irrigation programming [18, 19, 20, 21], This chapter aims to (i) reporting most latest and common techniques for the use of UAV imagery in precision agriculture for crop fields monitoring. (ii) discovering most important UAV system technologies and effective methods for data processing that can be utilized in order to exploit the agricultural data acquired by UAV system, and (iii) using thermal infrared imagery taken by employing UAV anticipate the changes in plant water use because of irrigation management. This method can be a decision-making aid in the future for the successful irrigation management of crops.

1 Remote Sensing Using an Unmanned Aerial Vehicle

Remotely piloted aircraft, also known as unmanned aerial vehicles or drones, has recently gained remote sensing popularity. In the last 35 years, UAV has been widely used for tracking cultivated fields and offering efficient precision agriculture solutions [22]. Images at different wavelengths can be used to track various crop and vegetation parameters by using remote sensing. Previously, remote sensing was frequently dependent on satellite images [23, 24]. Precision agriculture and smart farming are two domains where several reviews exist about the application of the UAV. The majority of the articles concentrate on the various forms of applications that UAV can have in environmental monitoring in general [25].

1.1 Unmanned Aerial Vehicle Surveillance of Crops

Unmanned aerial vehicle systems have proven to be successfully applied in various precision agriculture applications to the present, including various applications in herbicide, water shortage recognition, disease identification, etc. For calculating yield to solve the problem that found and optimize production, many decisions can be taken depending on the information gathered by the UAV. The following are the popular usage of unmanned aerial vehicles utilized in precision agriculture that has been documented in the literature:

- Weed mapping and management [26, 27]
- Plants growth monitoring and yield calculation [28, 29, 30]
- Disease identification and vegetation quality control [31, 32]
- Irrigation management [33, 34]
- Spraying by the Corps [35, 36]

Through visual analysis of images is used in many remote sensing operations, the Global Positioning System (GPS) technologies are widely applied to help in the localization and georeferencing of images obtained by remote sensing, which allocate reality accords to a picture. Remote sensing information is such a valuable source of data's environment, and also with other datasets, it is reviewed and incorporated frequently through importation into GIS. In their study, experts in remote sensing generally utilize GIS in conjunction with remote sensing imagery, GPS, and different type of spatial data. Other sensors, such as RADAR and LIDAR produce, calculate electromagnetic radiation returns [37]. The sensor's distance from the detected target is less than a given threshold, normally in the range of a few meters, by using remote sensing techniques [38].

1.2 Unmanned Aerial Vehicle System Technologies

In precision agriculture, main elements such as the following are often used in remotely sensed aircraft [39].

1.2.1 Ground Control Station (GCS)

The UAV is controlled and tracked specifically due to a computing devise termed ground control station connected to UAV's system control. The information relevant to the UAV's flight is tracked by the GCS. The consumer may obtain data relating to the aircraft's operation and data collected by the sensors that assist the flight. Moreover, GCS includes tools needed for UAV's information to be processed and the data that the device operator needs for field tracking to be extracted.

1.2.2 UAV Control System (UAV CS)

The UAV Control System controls the UAV, which could be as obvious as the control from a remote or a constructed GPS. The system of flight control and/or the system of autopilot that controls UAV movements are included in the UAV CS. For proper functioning of UAV, collecting information and data processing are managed from the autopilot or flight control system due to control system. Processing data from sensors by the control system fix any issues that may occur and connect with the GCC wirelessly and simultaneously transmit and receive information.

1.2.3 Data Collection Using Sensors

The data needed is captured by cameras, defined as sensors for collection information. There are various methods for gathering data and the technology used. When UAVs aren't used for data collection but are instead used for something else, for example, splattering, the sensors have been replaced with the appropriate constituents.

1.3 Unmanned Aerial Vehicle advantages

In addition to the most crucial UAV technologies above and being used in precision agriculture, the size of the UAVs is used for classification. Nevertheless, since it includes more parameters influencing the application of UAV and their efficiency in precision agriculture, the categorization used is more common. Most of recent precision agriculture research has used multirotor for UAVs. This would be mainly because the field that is taking into account in most applications isn't very extensive. Consequently, a high rapidity of UAVs and good capacity for covering vast regions in a few flights, such as fixed-wing UAVs, are unnecessary. As a result, aircraft with rotary wing are favored due to the following benefits:

Ability to move around
Simple to operate

Slower rates
Costs are relatively minimal

These advantages make it easier to collect data from crops using photography, which is how UAVs are often used in vegetation tracking. Aircraft fixed wing has opted where the location of the control regions seems to be wide since they can follow-up the whole area in a limited amount of time [39].

2 Precision Agriculture Using Unmanned Aerial Vehicle

Equipped with advanced sensors, unmanned aerial vehicles become strong systems for sensing, which is a supplement based on the Internet of Things methods. The sensor's function is to gather photographs with high resolution in space and time. Several various forms of sensors may be used based on crop parameters that need to be monitored [40]. Nonetheless, many constraints have been posted on the sensors due to the need for low payload capability and small platforms. The sensors must satisfy three key requirements: lightweight, low energy consumption, and small size. All of the preceding would, of course, be paired with the capability to take high-quality photographs.

Current trading sensors that meet the precedent requirements and are applied for precision agriculture primarily fall into one of four classifications:

- Visible light sensors
- Hyperspectral sensors
- Thermal sensors
- Multispectral sensors

Other various types of sensors are known as light detection and ranging sensors in the literature. Laser scanners can be employed in addition to the ones previously mentioned, also named **LiDAR**. Laser scanners are well-known technique, which are widely applied in the science of the environment, but they are frequently employed for scanning on the ground. Since 1994, scanning laser airborne has been used [41].

2.1 Unmanned Aerial Vehicle Data Acquisition

The information acquired with employing these sensors makes it possible to apply it to assess moisture in soil, control plant biomass, crop quality, and more essential vegetation specifics at different stages of development. The most important varieties of sensors utilized are:

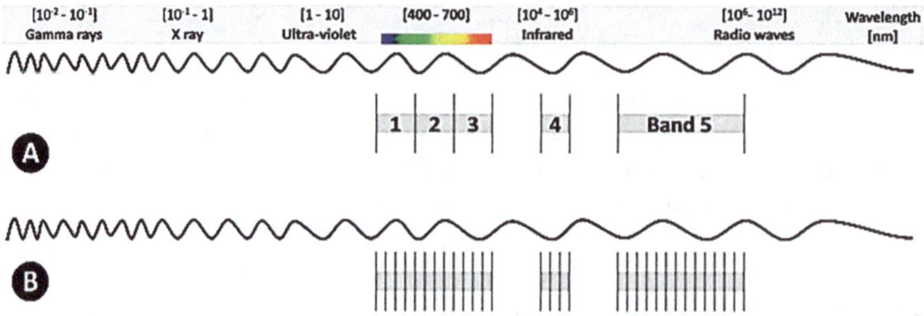

Fig. 1 (**a**) A multispectral example of five wide bands, (**b**) a hyperspectral example carries various narrow bands [42]

2.1.1 Visible Light Sensors (RGB)

They have the ability to capture high-quality photographs. The photographs can be taken in various conditions, with both sunny and cloudy conditions. Still, a strict timescale dependent on climate conditions is needed in order to prevent under or overexposure of the image.

2.1.2 Multispectral Imaging Sensors

Imaging sensors that are multispectral or hyperspectral are supplied in UAVs can collect data on absorption and reflection in the spectral range for the crop through many bands (Fig. 1). Data from the spectral may be very useful for evaluating various biological and physical features of crops, despite their higher costs [40].

2.1.3 Thermal Infrared Sensors

Infrared thermal sensors have to capture the temperature of the surfaces. Thermal cameras receive infrared energy by using infrared sensors and an optical lens. The temperature is higher than zero (273.3°C/459°F), and all objects send out infrared rays at various wavelengths LWIR and MWIR bands in proportion to the corresponding temperature. As a result, thermal imaging cameras concentrate and sense radiation in the same wavelengths, converting them into gray scale images to represent heat. Several sensed thermal imaging can produce image color. This kind of sensor is only found in a few applications for irrigation management. Consequently, infrared imaging sensors are rarely employed in precision agriculture implementations of UAV focused on tracking crop features. The spectral and spatial data related to each acquiring method were measure by applying a categorization based on a rating with a bullet (1–3), as shown in Table 1.

Table 1 The important distinction exists in hyperspectral and multispectral imaging, spectroscopy, and RGB imaging (merging viewpoints Wu and Sun [43] and Boreman [44]

	Spectral information	Spatial information
Hyperspectral imaging	•••	•••
Multispectral imaging	••	•••
Spectroscopy	•••	•
RGB imagery	•	•••

Table 2 Monitoring crops features with UAVs

	Crop features
Vegetation	Biomass [45, 46] Nitrogen status [45, 47, 48, 46] Moisture content [49, 48] Vegetation color [50, 51] Chlorophyll spectral behavior [33, 47] Temperature measurement [33, 52] Object's spatial position [53, 54] Vegetation index, various component's dimension, and shape [51, 55]
Soil	Humidity content [49, 56] Temperature measurement [57, 52] Electrical conductivity calculation [57]

2.2 UAS Data Processing

Various forms of the UAV data captured can be exploited to investigate various types of vegetation characteristics. The most shared characteristics that can be tracked using unmanned aerial vehicles for precision agriculture are the following in Table 2.

2.3 Techniques to Analyze UAV Imagery

Sensors for a specific purpose can help the unmanned aerial vehicle for obtaining information in different cultivated domains. The most frequent utilized image processing techniques to examine UAV imagery are the following:

2.3.1 Photogrammetry Techniques

Photogrammetric techniques may procedure the two-dimensional information and set up the geometrical liaison among the various images and objects, achieving 3-D models. This type of method has been exploited in order to extract 3-D digitized models of terrain [58, 59, 53] and also employed in orthophotos [60, 61].

2.3.2 Machine Learning Techniques

Using machine learning (ML) methods can be employed in order to procedure the information procured to forecasting and detection aim, with important effect in a variety field, as an example, healthcare systems [62, 61] marketing strategies [63], biology projects [64], and so on. This method can evaluate parameters concerning yield development, detect illness, or still identify objects in the images.

2.3.3 Vegetation Indices Calculation

Vegetation Index (VIs) calculation can utilize various mathematical combinations of a minimum of two bands spectral in the electromagnetic spectrum, intended for optimizing the crop specifics part although reducing the outward confusion element. They can provide trustworthy space-based and temporary data about the agricultural data of yield controlled.

3 The Thermal-Based Crop Stress Sensing

Crop irrigation technology has become an important domain where UAVs could be used in precision agriculture. Crop irrigation consumes 70% of the world's water [65, 66].

Emphasizing precision irrigation techniques' importance, precision irrigation methods may help to improve the quality of water by ensuring the efficient exploitation of the resource: in the proper places, at the appropriate times, with the correct quantities. Precision agriculture application divides the field into different sections of irrigation areas to better manage water. Unmanned aerial vehicles with suitable sensor variations may be used for detecting areas where vegetation requires much water. The precedent methods simultaneously enable customized maps that show the soil morphology to be developed so that the irrigation of every crop can be planned more efficiently separately.

The use of infrared sensors for sensing temperature or imaging displays an important ability to calculate CWSI and valuate irrigation plant water status [18, 19, 20, 21, 67].

The upper baseline and the water baseline were recorded as nonwater stressed baseline for CWSI measurement. [68].

3.1 Unmanned Aerial Vehicle and Thermal Imaging

The thermal pictures captured with the UAV helps for vegetation's temperature measurement by measuring the temperature of the pure plant pixels and averaging

Table 3 The thermal camera's main characteristics (Thermal Camera Tau 2324) [33]

	Camera characteristics
Range's scene	From −25°C to 135°C
Detecting device	Vanadium Oxide (VOx) microbolometer detector
Video resolution	324×256 pixels
Infrared imaging lens	9 mm f/1.25
Sensitivity to temperature	<50 mK
View wide field	48°×37°
Rates of full frame	30/60 Hz
Pixel-pitch	25 mm

them using a thermal camera with specific characteristics (Table 3). Therefore, a segmentation algorithm ranked R-based can be determined by histogram examination of each thermal picture. The full width at half limit (FWHM) allows pure vegetation pixel to be obtained from the thermal picture. The FWHM rule enables the pixels to have vegetation with a high possibility of absolute purity and be identified, as stated [69, 70].

3.1.1 Water Stress Cumulative Integration

The evaluation of the water status in the vegetation cover is dependent on the difference among the temperature average in the crop and the current temperature of air through the time of flight **ΔT**, which is determined with the sensor of air temperature. Values **ΔT** are also utilized to calculate a cumulative integration of water stress in the vegetation degree throughout the season of irrigation. **ΔT** Values can also be applied during the irrigation season to calculate a cumulative integration of water tension in the plant. In addition, the difference among **ΔT** and a canopy nonstressed for a particular sampling date gives the difference exist mainly in temperature of canopy or the level of stress. K. C. DeJonge [71] adapts this indicator.

Water stress integral WSI helps integrate the water stress effect during the entire day, initially suggested by Myers [72]. To measure water potential in predawn leaf for the whole irrigation season, we will use the total integral of water stress as determined by canopy temperature differences according to the following equation (1):

$$WSI\left({}^{\circ}C\ day\right) = \sum_{i=0}^{i=t}\left(\mathbf{\Delta T}_{i,i+1} - \mathbf{C}_{i,i+1}\right)n \tag{1}$$

t: the number of ΔT calculations, according to the number of flights conducted.
$\mathbf{\Delta T}_{\mathbf{i,i+1}}$: average of ΔT for any interval i, i+1.
$\mathbf{C}_{i,i+1}$: non-water-stressed values **ΔT** for any time frame i, i+1.
n: days number in the interval.
C: nonwater-stressed baseline value (**NWSB**) [68].

3.1.2 Measurement of the Crop Water Stress Indicator

Though irrigation programs, sensing the temperature of canopy employing infrared thermal sensors or imagery has revealed a big potentiality for determining the crop water stress index and identifying water status in the vegetation [17]. Equation 2 is using to measure the **CWSI** for the plant:

$$\mathbf{C}_i = \mathbf{a} + \mathbf{bVPD}_i \tag{2}$$

VPD$_i$: The total vapor pressure deficiency (kPa) at flight time through the ith day of calculation.
a, **b**: variables got for every species at particular environment conditions [17].

In the case study of Quebrajo et al. [33], they collect the vapor pressure deficiency and the temperature of air values and at a height of 2 meter in a weather station adjacent to the Andalusia government's Agroclimatic Information Network. To calculate the value of **CWSI** of the ith sampling day, the **NWSB** from Idso [68] can also be used as shown in Eq. 3:

$$\mathbf{CWSI} = \frac{\boldsymbol{\Delta}\mathbf{T}_{i-\mathbf{C}_{i,i+1}}}{\boldsymbol{\Delta}\mathbf{T}_{dry,i} - \mathbf{C}_{i,i+1}} \tag{3}$$

$\boldsymbol{\Delta}\mathbf{T}_{dry}$: The maximum value ΔT, according to a nontranspiring canopy.
$\boldsymbol{\Delta}\mathbf{T}_{dry}$: The maximum value ΔT, following nontranspiring canopy.

Based on this study [33], the sugar beet leaf temperature value can meet 8°C above air temperature; thus, $\boldsymbol{\Delta}\mathbf{T}_{dry,i}$ =8 C can be used as a constant. For other herbaceous crop types, similar values of $\boldsymbol{\Delta}\mathbf{T}_{dry,i}$ were found [69].

Crop water stress indicator's calculation is based on aerial thermal pictures provides an effective decision-making instrument in terms of adding value to soil humidity data during irrigation programs [15].

CWSI employment can reveal the variation between plants grown in different soil types and between well-irrigated plants and plants being sensitive to water deficiency for irrigation programming. Growth features related to humid baseline are specific to Mediterranean and geographical variations and the irrigation control establishment. Threshold values are based on alliances with other physiological variations in vegetation cover. It must be determined [33].

4 Conclusion

In this chapter, we present the most often used monitoring detectors for the vegetation, and we highlight the techniques of processing employing UAV imaging and the important exploitation of UAV. Using the remotely sensed aircraft systems in irrigation management can be more similar to a normative process, consisting of applying remote sensors: thermal and/or multispectral in order to follow-up the necessity of water in various crop portions. Furnished with thermal sensors, the UAV can identify possible shortages through irrigation in various spots. Such data is converted via photogrammetry methods to a high-quality mapping, noting the regions in stress. In that case, a map is created with VI application when multispectral imaging is employed. As a result, a map like this will encourage farmers to apply variable-rate water irrigation control. The differences in the temperature of canopy-air and the CWSI values obtained from an infrared thermal picture taken employing a remotely sensed aircraft were detected to determine changes in plant water usage generating through precision irrigation.

References

1. A. H. E. Nahry, R. R. Ali, and A. A. E. Baroudy, "An approach for precision farming under pivot irrigation system using remote sensing and GIS techniques," *Agricultural Water Management*, vol. 98, no. 4, pp. 517–531, 2011, doi: https://doi.org/10.1016/j.agwat.2010.09.012.
2. L. Haghverdi, F. Buettner, and F. J. Theis, "Diffusion maps for high-dimensional single-cell analysis of differentiation data," *Bioinformatics*, vol. 31, no. 18, pp. 2989–2998, 2015, doi: https://doi.org/10.1093/bioinformatics/btv325.
3. L. Zhang, M. L. Clarke, M. D. Steven, and K. W. Jaggard, "Spatial patterns of wilting in sugar beet as an indicator for precision irrigation," *Precision Agriculture*, vol. 12, no. 2, pp. 296–316, 2011, doi: https://doi.org/10.1007/s11119-010-9177-2.
4. I. PÔças *et al.*, "Satellite-based evapotranspiration of a super-intensive olive orchard: Application of METRIC algorithms," *Biosystems Engineering*, vol. 128, pp. 69–81, 2014, doi: https://doi.org/10.1016/j.biosystemseng.2014.06.019.
5. É. L. Simms and H. Ward, "Multisensor NDVI-based monitoring of the tundra-taiga interface (Mealy Mountains, Labrador, Canada)," *Remote Sensing*, vol. 5, no. 3, pp. 1066–1090, 2013, doi: https://doi.org/10.3390/rs5031066.
6. J. Senthilnath, M. Kandukuri, A. Dokania, and K. N. Ramesh, "Application of UAV imaging platform for vegetation analysis based on spectral-spatial methods," *Computers and Electronics in Agriculture*, vol. 140, pp. 8–24, 2017, doi: https://doi.org/10.1016/j.compag.2017.05.027.
7. M. Pérez-Ortiz, J. M. Peña, P. A. Gutiérrez, J. Torres-Sánchez, C. Hervás-Martínez, and F. López-Granados, "A semi-supervised system for weed mapping in sunflower crops using unmanned aerial vehicles and a crop row detection method," *Applied Soft Computing Journal*, vol. 37, pp. 533–544, 2015, doi: https://doi.org/10.1016/j.asoc.2015.08.027.
8. Jay, S. et al., "Exploiting the centimeter resolution of UAV multispectral imagery to improve remote-sensing estimates of canopy structure and biochemistry in sugar beet crops," *Remote Sensing of Environment.*, 231, September, 1, 2019, https://doi.org/10.1016/j.rse.2018.09.011.
9. A. Hall, J. Louis, and D. Lamb, "A method for extracting detailed information from high resolution multispectral images of vineyards," *Proceedings of the 6th International Conference on GeoComputation*, January 2001, p. ISBN 1864995637.
10. F. López-Granados, J. Torres-Sánchez, A. Serrano-Pérez, A. I. de Castro, F. J. Mesas-Carrascosa, and J. M. Peña, "Early season weed mapping in sunflower using UAV technology: Variability of herbicide treatment maps against weed thresholds," *Precision Agriculture*, vol. 17, no. 2, pp. 183–199, 2016, doi: https://doi.org/10.1007/s11119-015-9415-8.
11. M. N. Reza, I. S. Na, S. W. Baek, and K. H. Lee, "Rice yield estimation based on K-means clustering with graph-cut segmentation using low-altitude UAV images," *Biosystems Engineering*, vol. 177, no. 2018, pp. 109–121, 2019, https://doi.org/10.1016/j.biosystemseng.2018.09.014.

12. J. R. Ducati, M. G. Bombassaro, and J. M. G. Fachel, "Classifying vineyards from satellite images: A case study on burgundy's côte d'or," *Journal International des Sciences de la Vigne et du Vin*, 48, 4, pp. 247–260, 2014, doi: https://doi.org/10.20870/oeno-one.2014.48.4.1693.
13. J. L. Hatfield, A. A. Gitelson, J. S. Schepers, and C. L. Walthall, "Application of spectral remote sensing for agronomic decisions," *Agronomy Journal*, vol. 100, no. 3 SUPPL, pp. 117–131, 2008, doi: https://doi.org/10.2134/agronj2006.0370c.
14. E. R. Hunt, P. C. Doraiswamy, J. E. McMurtrey, C. S. T. Daughtry, E. M. Perry, and B. Akhmedov, "A visible band index for remote sensing leaf chlorophyll content at the Canopy scale," *International Journal of Applied Earth Observation and Geoinformation*, vol. 21, no. 1, pp. 103–112, 2012, doi: https://doi.org/10.1016/j.jag.2012.07.020.
15. J. Bellvert, P. J. Zarco-Tejada, J. Marsal, J. Girona, V. González-Dugo, and E. Fereres, "Vineyard irrigation scheduling based on airborne thermal imagery and water potential thresholds," *Australian Journal of Grape and Wine Research*, vol. 22, no. 2, pp. 307–315, 2016, doi: https://doi.org/10.1111/ajgw.12173.
16. M. Greco, A. Chiappetta, L. Bruno, and M. B. Bitonti, "In Posidonia oceanica cadmium induces changes in DNA methylation and chromatin patterning," *Journal of Experimental Botany*, vol. 63, no. 2, pp. 695–709, 2012, doi: https://doi.org/10.1093/jxb/err313.
17. R. D. Jackson, S. B. Idso, R. J. Reginato, and P. J. Pinter, "Canopy temperature as a crop water stress indicator," *Water Resources Research*, vol. 17, no. 4, pp. 1133–1138, 1981, doi: https://doi.org/10.1029/WR017i004p01133.
18. M. P. González-Dugo, M. S. Moran, L. Mateos, and R. Bryant, "Canopy temperature variability as an indicator of crop water stress severity," *Irrigation Science*, pp. 1–8, 2006, doi: https://doi.org/10.1007/s00271-005-0023-7.
19. J. O. Payero, D. Tarkalson, and S. Irmak, "Corn yield response to different irrigation depths with subsurface drip irrigation," *Examining Conflu. Environ. Water Concerns - Proc. World Environ. Water Resour. Congr. 2006*, pp. 1–18, 2007, doi: https://doi.org/10.1061/40856(200)274.
20. M. Möller *et al.*, "Use of thermal and visible imagery for estimating crop water status of irrigated grapevine," *Journal of Experimental Botany*, vol. 58, no. 4, pp. 827–838, 2007, doi: https://doi.org/10.1093/jxb/erl115.
21. L. Testi, D. A. Goldhamer, F. Iniesta, and M. Salinas, "Crop water stress index is a sensitive water stress indicator in pistachio trees," *Irrigation Science*, vol. 26, no. 5, pp. 395–405, 2008, doi: https://doi.org/10.1007/s00271-008-0104-5.
22. D. J. Mulla, "Twenty five years of remote sensing in precision agriculture: Key advances and remaining knowledge gaps," *Biosystems Engineering*, vol. 114, no. 4, pp. 358–371, 2013, doi: https://doi.org/10.1016/j.biosystemseng.2012.08.009.
23. M. E. Bauer, "Identification of agricultural crops by computer processing of ERTS-MSS data," *LARS Technology Reports. Paper 20.*, 1973.
24. R. A. R. Mora André, Tiago M. A. Santos, Szymon Łukasik, João M.N. Silva, António Falcão, Jose Manuel Fonseca, "Land cover classification from multispectral data using computational intelligence tools: A comparative study," *Information*, 8, 4, 2017, doi: https://doi.org/10.3390/info8040147.
25. S. Manfreda *et al.*, "On the use of unmanned aerial systems for environmental monitoring," *Remote Sensing.*, 10, 4, 2018, doi: https://doi.org/10.3390/rs10040641.
26. M. D. Bah, A. Hafiane, and R. Canals, "Weeds detection in UAV imagery using SLIC and the hough transform," *Proceedings of the Seventh International Conference on Image Processing Theory, Tools and Applications - IPTA 2017.* 2018-January, pp. 1–6, 2018, https://doi.org/10.1109/IPTA.2017.8310102.
27. A. dos Santos Ferreira, D. Matte Freitas, G. Gonçalves da Silva, H. Pistori, and M. Theophilo Folhes, "Weed detection in soybean crops using ConvNets," *Computers and Electronics in Agriculture.*, vol. 143, no. February, pp. 314–324, 2017, doi: https://doi.org/10.1016/j.compag.2017.10.027.

28. J. Jung, M. Maeda, A. Chang, J. Landivar, J. Yeom, and J. McGinty, "Unmanned aerial system assisted framework for the selection of high yielding cotton genotypes," *Computers and Electronics in Agriculture.*, 152, June, pp. 74–81, 2018, doi: https://doi.org/10.1016/j.compag.2018.06.051.
29. L. Han, G. Yang, H. Yang, B. Xu, Z. Li, and X. Yang, "Clustering field-based maize phenotyping of plant-height growth and canopy spectral dynamics using a UAV remote-sensing approach," *Frontiers in Plant Science.*, vol. 871, no. November, pp. 1–18, 2018, doi: https://doi.org/10.3389/fpls.2018.01638.
30. I. Wahab, O. Hall, and M. Jirström, "Remote sensing of yields: Application of UAV imagery-derived ndvi for estimating maize vigor and yields in complex farming systems in Sub-Saharan Africa," *Drones*, vol. 2, no. 3, pp. 1–16, 2018, doi: https://doi.org/10.3390/drones2030028.
31. D. Montero and C. Rueda, "Detection of palm oil bud rot employing artificial vision," *IOP Conference Series: Materials Science and Engineering.*, 437, 1, 2018, doi: https://doi.org/10.1088/1757-899X/437/1/012004.
32. M. Kerkech, A. Hafiane, and R. Canals, "Deep leaning approach with colorimetric spaces and vegetation indices for vine diseases detection in UAV images," *Computers and Electronics in Agriculture.*, 155, October, pp. 237–243, 2018, doi: https://doi.org/10.1016/j.compag.2018.10.006.
33. L. Quebrajo, M. Perez-Ruiz, L. Pérez-Urrestarazu, G. Martínez, and G. Egea, "Linking thermal imaging and soil remote sensing to enhance irrigation management of sugar beet," *Biosystems Engineering*, vol. 165, pp. 77–87, 2018, doi: https://doi.org/10.1016/j.biosystemseng.2017.08.013.
34. C. Albornoz and L. F. Giraldo, "Trajectory design for efficient crop irrigation with a UAV," *2017 2017 IEEE 3rd Colombian Conference on Automatic Control. CCAC 2017 - Conference Proceedings.*, 2018-January, pp. 1–6, doi: https://doi.org/10.1109/CCAC.2017.8276401.
35. X. Xue, Y. Lan, Z. Sun, C. Chang, and W. C. Hoffmann, "Develop an unmanned aerial vehicle based automatic aerial spraying system," *Computers and Electronics in Agriculture*, vol. 128, pp. 58–66, 2016, doi: https://doi.org/10.1016/j.compag.2016.07.022.
36. P. Garre and A. Harish, "Autonomous Agricultural Pesticide Spraying UAV," *IOP Conference Series: Materials Science and Engineering.*, 455, 1, 2018, doi: https://doi.org/10.1088/1757-899X/455/1/012030.
37. R. Keranen, K. Kolvoord, "Making spatial decisions using ArcGIS Pro: A workbook. By Kathryn Keranen and Robert Kolvoord, Esri Press, 2017; 376 Pages. Price $69.99, ISBN 9781589484849," *ISPRS International Journal of Geo-Information*, 6, 9, p. 279, 2016, doi: https://doi.org/10.3390/ijgi6090279.
38. R. A. Viscarra Rossel, V. I. Adamchuk, K. A. Sudduth, N. J. McKenzie, and C. Lobsey, *Proximal soil sensing: An effective approach for soil measurements in space and time*, 113. Elsevier Inc., 2011.
39. S. G. Gupta, M. Ghonge, and P. M. Jawandhiya, "Review of Unmanned Aircraft System (UAS)," *SSRN Electronic Journal.*, April, 2013, doi: https://doi.org/10.2139/ssrn.3451039.
40. G. Yang *et al.*, "Unmanned aerial vehicle remote sensing for field-based crop phenotyping: Current status and perspectives," *Frontiers in Plant Science.*, 8, June, 2017, doi: https://doi.org/10.3389/fpls.2017.01111.
41. J. Gajardo, M. García, and D. Riaño, *Applications of airborne laser scanning in forest fuel assessment and fire prevention*. 2014.
42. G. L. Anderson, J. H. Everitt, A. J. Richardson, and D. E. Escobar, "Using satellite data to map false broomweed (Ericameria austrotexana) infestations on south Texas rangelands," *Weed Technology*, 7, 4, pp. 865–871, 1993, https://doi.org/10.1017/s0890037x00037908.
43. D. Wu and D.-W. Sun, "Advanced applications of hyperspectral imaging technology for food quality and safety analysis and assessment: A review – Part I: Fundamentals," *Innovative Food Science and Emerging Technologies*, vol. 19, pp. 1–14, 2013, doi: https://doi.org/10.1016/j.ifset.2013.04.014.
44. G. D. Boreman, "Classification of imaging spectrometers for remote sensing applications," *Optical Engineering.*, 44, 1, p. 013602, 2005, doi: https://doi.org/10.1117/1.1813441.

45. R. Näsi *et al.*, "Estimating biomass and nitrogen amount of barley and grass using UAV and aircraft based spectral and photogrammetric 3D features," *Remote Sensing*, vol. 10, no. 7, pp. 1–32, 2018, doi: https://doi.org/10.3390/rs10071082.
46. M. Maimaitijiang *et al.*, "Unmanned Aerial System (UAS)-based phenotyping of soybean using multi-sensor data fusion and extreme learning machine," *ISPRS Journal of Photogrammetry and Remote Sensing*, vol. 134, pp. 43–58, 2017, doi: https://doi.org/10.1016/j.isprsjprs.2017.10.011.
47. C. Ballester, J. Hornbuckle, J. Brinkhoff, J. Smith, and W. Quayle, "Assessment of in-season cotton nitrogen status and lint yield prediction from unmanned aerial system imagery," *Remote Sensing*, 9, 11, pp. 1–18, 2017, doi: https://doi.org/10.3390/rs9111149.
48. S. Bhandari *et al.*, "Lessons Learned from UAV-Based Remote Sensing for Precision Agriculture $_*$," *2018 International Conference on Unmanned Aircraft Systems. ICUAS 2018*, pp. 458–467, https://doi.org/10.1109/ICUAS.2018.8453445.
49. J. Jorge, M. Vallbé, and J. A. Soler, "Detection of irrigation inhomogeneities in an olive grove using the NDRE vegetation index obtained from UAV images," *European Journal of Remote Sensing*, 52, 1, pp. 169–177, 2019, https://doi.org/10.1080/22797254.2019.1572459.
50. L. Wan et al., Combining UAV-based vegetation indices and image classification to estimate flower number in oilseed rape, *Remote Sensing***10**(9), 2018, https://doi.org/10.3390/rs10091484.
51. S. Varela *et al.*, "Early-season stand count determination in Corn via integration of imagery from unmanned aerial systems (UAS) and supervised learning techniques," *Remote Sensing.*, 10, 2, 2018, doi: https://doi.org/10.3390/rs10020343.
52. S. Pascuzzi, A. S. Anifantis, V. Cimino, and F. Santoro, "Unmanned aerial vehicle used for remote sensing on an apulian farm in southern Italy," *Engineering for Rural Development.*, 17, pp. 149–154, 2018, doi: https://doi.org/10.22616/ERDev2018.17.N175.
53. A. I. de Castro, J. Torres-Sánchez, J. M. Peña, F. M. Jiménez-Brenes, O. Csillik, and F. López-Granados, "An automatic random forest-OBIA algorithm for early weed mapping between and within crop rows using UAV imagery," *Remote Sensing*, 10, 2, pp. 1–21, 2018, https://doi.org/10.3390/rs10020285.
54. J. Gao *et al.*, "Fusion of pixel and object-based features for weed mapping using unmanned aerial vehicle imagery," *International Journal of Applied Earth Observation and Geoinformation.*, 67, December 2017, pp. 43–53, 2018, doi: https://doi.org/10.1016/j.jag.2017.12.012.
55. M. de J Marcial-Pablo, A. Gonzalez-Sanchez, S. I. Jimenez-Jimenez, R. E. Ontiveros-Capurata, and W. Ojeda-Bustamante, "Estimation of vegetation fraction using RGB and multispectral images from UAV," *International Journal of Remote Sensing*, vol. 40, no. 2, pp. 420–438, 2019, doi: https://doi.org/10.1080/01431161.2018.1528017.
56. R. Sobayo, H. H. Wu, R. Ray, and L. Qian, "Integration of convolutional neural network and thermal images into soil moisture estimation," *Proc. - 2018 1st Int. Conf. Data Intell. Secur. ICDIS 2018*, pp. 207–210, 2018, doi: https://doi.org/10.1109/ICDIS.2018.00041.
57. K. Křížová, J. Haberle, M. Kroulík, J. Kumhálová, and J. Lukáš, "Assessment of soil electrical conductivity using remotely sensed thermal data," *Agronomy Research.*, 16, 3, pp. 784–793, 2018, doi: https://doi.org/10.15159/AR.18.111.
58. M. G. Ziliani, S. D. Parkes, I. Hoteit, and M. F. McCabe, "Intra-season crop height variability at commercial farm scales using a fixed-wing UAV," *Remote Sensing*, vol. 10, no. 12, pp. 1–25, 2018, doi: https://doi.org/10.3390/rs10122007.
59. J. Torres-Sánchez *et al.*, "Mapping the 3D structure of almond trees using UAV acquired photogrammetric point clouds and object-based image analysis," *Biosystems Engineering*, vol. 176, pp. 172–184, 2018, doi: https://doi.org/10.1016/j.biosystemseng.2018.10.018.
60. I. Sa *et al.*, "WeedMap: A large-scale semantic weed mapping framework using aerial multispectral imaging and deep neural network for precision farming," *arXiv*, pp. 1–25, 2018.
61. S. Marino and A. Alvino, "Detection of homogeneous wheat areas using multi-temporal UAS images and ground truth data analyzed by cluster analysis," *European Journal of Remote Sensing*, 51, 1, pp. 266–275, 2018, doi: https://doi.org/10.1080/22797254.2017.1422280.

62. D. C. Tsouros *et al.*, "Automated collagen proportional area extraction in liver biopsy images using a novel classification via clustering algorithm," *Proceedings of the IEEE Symposium on Computer-Based Medical Systems.*, 2017-June, pp. 30–34, 2017, https://doi.org/10.1109/CBMS.2017.99.
63. D. Cui and D. Curry, "Prediction in marketing using the support vector machine," *Marketing Science*, vol. 24, no. 4, pp. 595–615, 2005, doi: https://doi.org/10.1287/mksc.1050.0123.
64. A. L. Tarca, V. J. Carey, X. Wen Chen, R. Romero, and S. Drăghici, "Machine learning and its applications to biology.," *PLoS Computational Biology.*, 3, 6, 2007, https://doi.org/10.1371/journal.pcbi.0030116.
65. K. Chartzoulakis and M. Bertaki, "Sustainable water management in agriculture under climate change," *Agriculture and Agricultural Science Procedia*, 4, pp. 88–98, 2015, doi: https://doi.org/10.1016/j.aaspro.2015.03.011.
66. P. Saccon, "Water for agriculture, irrigation management," *Applied Soil Ecology.*, 123, November 2017, pp. 793–796, 2018, doi: https://doi.org/10.1016/j.apsoil.2017.10.037.
67. S. Taghvaeian, L. Comas, K. C. DeJonge, and T. J. Trout, "Conventional and simplified canopy temperature indices predict water stress in sunflower," *Agricultural Water Management*, 144, pp. 69–80, 2014, doi: https://doi.org/10.1016/j.agwat.2014.06.003.
68. S. B. Idso, "Non-water-stressed baselines: A key to measuring and interpreting plant water stress," *Agricultural Meteorology*, 27, 1–2, pp. 59–70, 1982, https://doi.org/10.1016/0002-1571(82)90020-6.
69. R. Rud *et al.*, "Crop water stress index derived from multi-year ground and aerial thermal images as an indicator of potato water status," *Precision Agriculture*, 15, 3, pp. 273–289, 2014, doi: https://doi.org/10.1007/s11119-014-9351-z.
70. J. Käthner, A. Ben-Gal, R. Gebbers, A. Peeters, W. B. Herppich, and M. Zude-Sasse, "Evaluating spatially resolved influence of soil and tree water status on quality of European plum grown in semi-humid climate," *Frontiers in Plant Science.*, 8, June, pp. 1–10, 2017, doi: https://doi.org/10.3389/fpls.2017.01053.
71. K. C. DeJonge, S. Taghvaeian, T. J. Trout, and L. H. Comas, "Comparison of canopy temperature-based water stress indices for maize," *Agricultural Water Management*, vol. 156, pp. 51–62, 2015, doi: https://doi.org/10.1016/j.agwat.2015.03.023.
72. Myers, B. J. (1988). Water stress integral-a link between shortterm stress and long-term growth, pp. 68–70, 1988.

Smart Healthcare Using Evolutionary Algorithm

Deepti Saraswat and Manik Lal Das

1 Introduction

Recently, healthcare domain is increasingly spreading its footsteps towards technological advancement in leveraging smart system to speed up health diagnostics and treatment. IoT in healthcare provides intelligent services for medical automation and monitoring of health in different environments (hospitals, workplaces, home etc.), thus allowing significant reduction of physician visit costs and enhanced quality care of the patient. Health services are a major concern in most developing countries because of the continued rising costs of precious non-consumable human resources involved. Healthcare implementation has big impact on digital transformation of smart cities which provides robust platform for planning, remote information processing, supervision and communication.

Internet of Medical Things (IoMT) has played an important role in unfolding smart health [12]. IoMT is conceptualized through sensor hardware, gateway and a wireless network that accredits communication and access of information to the users. IoT technologies have layered the viability in the development of a number of smarter solutions viz. paediatric care, elderly care, chronic disease monitoring, supervision of epidemic disease, medical cyber-physical management systems, supervision and regulation of private health [17]. IoT technology has removed

D. Saraswat (✉)
Institute of Technology, Nirma University, Ahmedabad, Gujarat, India

DA-IICT, Gandhinagar, Gujarat, India
e-mail: 201921013@daiict.ac.in

M. L. Das
DA-IICT, Gandhinagar, Gujarat, India
e-mail: maniklal_das@daiict.ac.in

M. Azrour et al. (eds.), *IoT and Smart Devices for Sustainable Environment*,
EAI/Springer Innovations in Communication and Computing,
https://doi.org/10.1007/978-3-030-90083-0_12

restrictions and limitations to hospital admitted patients as well as for self-care-takers at homes or any self-reliant care-taking centres. Wearable sensors, embedded with intelligence attached to a patient, can monitor various health status parameters. Complex data analysis through numerical and adaptive techniques such as artificial intelligence (AI) and machine learning/deep learning (ML/DL) is performed on the physiological data received from sensors, wearable through a network in a centralized system. The data after analysis is distributed to computing platform and resourced to comprehend the status of the concerned patients, and an advice or a diagnosis is stipulated accordingly through SMS or smartphone applications. Edge/fog computing in IoMT offers advantage in terms of fast data handling and processing, communication and storage on edge servers that are physically closed to the edge of the networks and improve the quality of service (QoS) for application that requires intensive computing, privacy, mobility support, scalability and low latency [1]. The former effectively determine accurate analysis of all disease type, real-time data processing with decision-making to cater with emergency situation of patients. Recently, smartphones are making their way not only as gateways [2] but also in the fields of edge computing. Cloud-based storage helps extract the historical statistics of patients related to the particular disease. The data is used for further offline analysis and decision-making process, maintaining privacy throughout.

Health-related issues are increasing dramatically and require immediate attention and reliability without compromising the quality of service. Currently, the health-care system is facing great problems due to an exponential increase in population as well as unpredictable spread of chronic diseases. Facility location management provides meaningful solution to implement low cost and scalable smart medical centres at various reachable locations of city. The search for optimal location(s) to facilitate in urban cities is one of the indispensable requirements in any spatial decision problems. Facility location models are utilized as a part of significant efforts to improve healthcare services as well and incorporate new features that represent real-life situations.

1.1 Motivation

As discussed above, an increase in population aiding with rapid spread of epidemic diseases has stressed huge demands for resources such as beds, nurses to doctor, critical care emergency units etc. in the hospital. The spread of epidemic diseases such as recent outbreak of COVID-19 pandemic heavily demands the latter for early detection and subsequent treatment. There is a constant increase in the traffic condition because of rapid urbanization, particularly in smart cities that impose challenges in enabling real-time medical diagnostics and follow-up treatment. People especially elderly ones find it difficult to incorporate daily movement in the congested traffic and large distances to be covered to hospital. The ambulance response time is recognised as an important measure of pre-hospital emergency medical services, and the situation becomes decisive because of to-and-fro movement of the former in heavy traffic conditions. A second concern is that many still do not have access to healthcare or to the fact that many patients live in semi-urban or dense areas with a shortage of medical healthcare professionals.

1.2 Our Contribution

The process of treatment in hospitals consists of various steps like locate the hospital, transportation time, admission and subsequent clinical treatment. The above is not feasible in all situations viz. critical/emergency conditions, age factor etc. as mentioned previously. Inspired from these facts, we propose a compact health facility in the form of smart medical centres (SMCs). The SMC will be established at various defined locations to aid accessibility to the patients of all ages.

The key highlights of the proposed research contributions are as follows:

- We propose SMC at various optimal locations in urban areas utilizing certain factors of the city where a patient can visit at any time to access healthcare in minimum time. *Smartbed*, an important constituent of SMC, will utilize IoT to sense various physiological parameters of the patient. An edge-enabled IoT network architecture at patient and doctor end obtains the health data, distributes work-flow resources, process and analyses and provides necessary acknowledgment/requisites from doctor end to the patient based on risk assessment and other factors.
- We propose an optimal planning scheme to predict most suitable locations for medical centres that covers maximum population of city. The algorithm produces the solution based on the following factors (objectives).

  1. Geographical information system (GIS) database containing information of roads, existing hospital/healthcare facilities and various other constraints (such as accident prone areas, highways, industrial areas, landfill sites etc.) in the studied area.
  2. Minimize the travel time to SMC and provide facilities and reach to maximum users. Avoid traffic-prone locations and prioritize areas with the minimum transportation cost.
  3. Maximize the accessibility to the users in population/narrow street areas with high density of population.

- We develop a genetic algorithm (GA) that considers random candidate site in population and obtains optimal points for establishment of SMC using single-objective criteria. Each objective function provides solution based on the required constraint. The algorithm derives the possible sites considering all the objectives as a multi-objective optimization (MOO).

The experimental results depict the uniqueness of proposed scheme as an effective solution for establishment of SMC. The overall scenario targets end-to-end diagnosis to the patients in a modern patient–doctor scenario.

1.3 Organization of the Chapter

The contributed research work is organized into five sections. Section 1 presents the motivation and our key contributions behind the research, Sect. 2 describes the related work in fields of IoT healthcare and various genetic algorithm-based optimization applications. Section 3 describes the architecture of the proposed smart medical centre (SMC). Section 4 describes the proposed scheme for prediction and establishment of the locations of SMC. Section 5 provides analysis and experimental results. Finally, Sect. 6 presents the conclusion.

2 Related Work

In the literature, many authors proposed solutions through addressing different optimization techniques to establish various human-centric facilities like hospitals, networks, ambulance locations etc. Also, a number of recent research works have been conducted in IoT-based smart healthcare that sense, share information and provide a wide range of services. A combination of edge, fog or cloud approach is basically used in conjunction with IoT in smart healthcare. Hakimi [14, 15] introduced the p-median problem while solving the optimal location of switching centres in communication networks. The author derives the optimal locations of switching centres to be at the vertex of communication network. It also discusses the utilization of Boolean functions for generating stable vertex coverings over vertices of a graph. Daskin and Dean [8] classify the facility location models used in healthcare applications using p-median or maximum covering model. The p-median model discussed in [25] minimizes the total travelling distance from all clients to their closest serving facilities. The research presents an overview of challenges and methods in the various areas of interest for the measurement of performance and healthcare facility location problem improvements. The authors in [6] discussed already established healthcare service in the states of USA with optimal locations with an aim to satisfy two objectives: minimization of hospital-patient distance and maximization of distance covered within a specified time. They have calculated travel distances and times using census block groups and road networks. The average travel time was within 3–6% of optimal value with a good service availability for 90–94% population within 30-min distance to hospitals. Geographical information system (GIS)-based location prediction technique is very popular for location determination. The authors in [31] describe the integer programming and GIS approach for optimization of location allocation problem of pharmacy warehouses. GIS and subsequently set covering, p-centre and p-median models were used to gather geographic information current and potential locations of pharmacy warehouses, hospitals and pharmacies and establishment, respectively. The authors in [24] discussed integrated methodology for healthcare location optimization using GIS, which is based on minimization of Euclidean distance between existing facilities and new facility using the Weber model and Weiszfeld's algorithm. They have successfully utilized GIS to locate real optimal location with respect to existing locations in the study area of Ijebu, Ogun State, Nigeria. The authors in [9, 10] determined the emergency support facility and cement plant locations using the Weber model and spatial data. They propose differential locations nearby to each other, which is most suitable for site development and decreases human casualties. Similarly, cement production facility location was determined using a common iterative methodology of the Weber model and spatial data coordinates.

Any healthcare facility planning introduces specific challenges that an algorithm is required to meet. The challenges are in terms of multiple objectives and criteria fulfilment. Recently, the location-oriented problems have been solved using various heuristic approaches. Genetic algorithm (GA) is the most used meta-heuristic method, which is widely used to derive location problems that consist of multiple objectives. GA methods are suitable for huge and polynomial spaces for global optimization problem [19]. The authors in [30] optimized the shape and location of sites using graphical representation of sites through multi-objective evolutionary algorithm. Here, a solution/site is encoded using a graphical representation operated on by a set of specifically designed evolutionary operations. A GA-based multi-objective modelling was used in [23] for urban park and open space (POS) location identification. The methodology considers four objective functions, POS provision near densely populated areas, polluted areas, highly disturbed areas and normal open spaces. The multi-objective model was implemented and executed using data sets gathered from the city of Dhaka. The authors in [5] compute the optimal location that is suitable for the establishment of new health facilities in the urban areas of Tehran. The authors have implemented solutions based on single-objective as well as multi-objective optimization. Four possible solutions for population of 40,000 in the city area of Tehran were derived using the said objectives.

The Internet of Things (IoT) connects several entities over Internet; these entities exchange data (acquired/read from surrounding environment) within the network to the process and analytic systems and govern a definite action. Numerous applications exist for IoT that covers a wide range such as manufacturing sector, logistics, agriculture, geology, healthcare, home automation, smart grid etc. In [22], the authors have proposed IoT-enabled intelligent framework for bio-gas detection in the city of Mohammedia, Morocco. The Arduino-based microcontroller framework monitors bio-gas and measures the level of various emitted gases. On the similar lines, the authors in [20] have developed a lightweight framework to monitor environmental (air and weather) parameters. The data is stored in a central computer, which is accessible for future utilization and decision-making. In [21], the authors have proposed a low-cost framework to detect the quality of water remotely. The framework interfaces various sensors to Arduino-based microcontroller and sends data to smart watch, smartphone via Wi-Fi/ Bluetooth connectivity and computer connected to the Internet. It measures parameters such as pH, conductivity, temperature and turbidity of water.

The IoT has propelled the healthcare applications in an extensive manner. A number of solutions exist and have been developed recently that integrate cloud and edge technologies in IoT architecture and provide highly effective responsiveness to healthcare systems. In [27], the authors have proposed a secure IoT-based health monitoring system where a micro-controller carries out necessary processing and sends the vital signals using Wi-Fi communication. It utilizes cloud as well as external server. In [26], the authors have presented wireless healthcare system for monitoring through Zigbee protocol. It consists of two parts viz. Arduino with Zigbee and Raspberry-Pi with Zigbee. Data communications occur between them. The system determines the transmission data related to measurement of body temperature on Zigbee wireless channel using two transceivers (Arduino and Raspberry-Pi). In [11, 18, 28], the authors have demonstrated IoT- and cloud-based technology to store, display real-time health data for remote monitoring, analysis as well as alert generation. A smart healthcare system based on edge cognitive computing is proposed in [7] that monitors the data, analyses and distributes computing resources as per health risk grade of patient. In [13], the authors focus on the issue related to detection of anomalies in health data parameters through implementation of the *HTM* algorithm running on edge nodes. The authors in [16] developed *AutoTriage* system for pre-scanning of COVID-19 and detection of fever and cyanosis without any physical contact.

As discussed above, Electronic Heath Record (EHR) data in IoT-oriented applications needs to be safeguard for secured access and distribution to authorised entities. In [3, 4], the authors have devised a secured and efficient protocol that utilizes five phases of authentication. The scheme is effective against various malicious activities such as mutual authentication, Denial-of-Service (DoS) attack, perfect forward secrecy, offline password guessing attack etc. The security analysis of the proposed scheme was verified using Scyther tool and is computationally faster for IoT applications.

In our research, we propose to establish SMCs at optimal locations derived using the genetic algorithm for various beneficiaries in a modern patient–doctor scenario. We also propose IoT-Edge architecture, an important constituent of SMC for information exchange which will provide end-to-end solution as in required for smart healthcare.

3 Proposed System Model

A smart medical centre (SMC) underlies a digitized networking infrastructure or a group of optimized clinical processes to provide a valuable service for better patient care, experience and operational efficiency, which was not possible or available earlier. It enables the use of various technologies such as healthcare smart devices, IoT, to fetch information dynamically, make things connected through internet, connect people with healthcare institutions and provide medications to the user by using artificial intelligence, cloud computing services and machine learning methods at edge devices in real time. Patient can dynamically access the various functionalities of the above system such as physical examination, scheduling of online appointments and patient interaction with doctor. Smart centres try to maximize satisfaction and flexibility for all classes of patients (younger and elderly) in normal and critical disease conditions.

In our research context, a smart medical centre consists of a smart bed fitted with IoT sensors and gadgets that monitors patients' physiological parameters. The data is relayed to the medical institutions, emergency services or doctor smart device through the communication network utilizing smart devices such as wearable, embedded controllers and BSN (body sensor network) at patient end.

3.1 Scenario

The proposed architecture can be used where the end users can be patients in hospital, patient at some remote location, sport playing people, people working in factory etc. who can share their health data. The smart healthcare system considers a patient and doctor scenario. Figure 1 depicts the architecture of the proposed smart medical centre. It consists of three important layers.

Patient Layer This layer consists of beneficiaries such as children, elderly, old age or patients themselves. *Smartbed*, an important entity, is interfaced with IoT sensors (physiological sensors) and IoT device (which is a mobile phone). The IoT device collects the key vitals of the body, i.e. heart rate, blood oxygen saturation (SpO_2), temperature and blood pressure from the sensors, and transmits them to computing

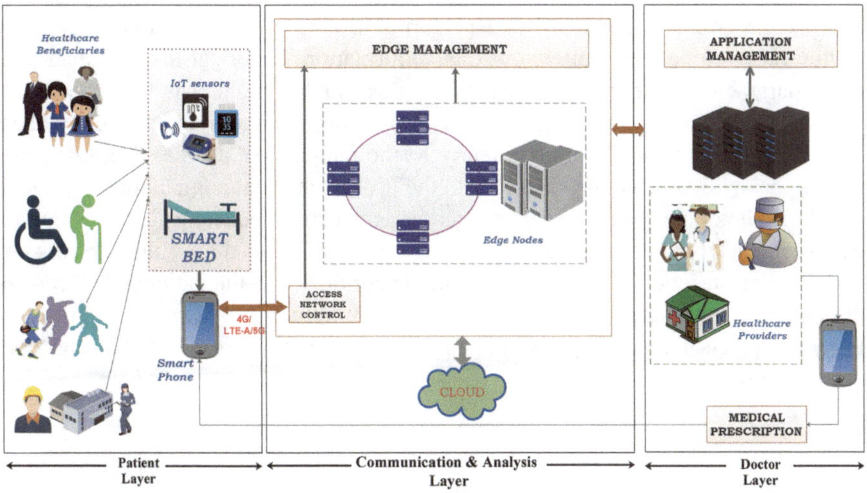

Fig. 1 Architecture of SMC

and analysis devices. Also, the mobile phone accepts the health analysis results from the IoT device (smartphone) via SMS or e-mail.

Communication and Analysis Layer This layer utilizes powerful edge/fog nodes that communicate with smart devices at patient end. The nodes receive, compute and analyse the physiological data and allocate computing assets of the beneficiaries based on different health levels and execute comprehensive remote medical diagnosis in real time through Internet networking (4G/LTE-A/5G). The cloud stores the basic information of the patient, medical history, registered information of doctor as well as prescriptions/requisites provided by the doctor.

Doctor Layer This layer consists of service providers such as doctors, trusted hospitals or emergency medical services (EMS) to view and monitor the vitals of patients through a monitoring system or an interactive application. The doctor can predict any unusual activity based on the data and prescribes the required medication/requisites to prevent the emergency case. The doctor layer also consists of a smartphone.

3.2 Workflow Between Patient and Doctor IoT Device with Edge Layers

The work flow is explained as follows. The patient locates the nearest SMC through an interactive mobile application. After the patient registers his/her identity at SMC, the smart bed will be allotted. Smart bed consists of different sensors to collect the patient health data. Smart bed initializes the IoT sensor system, collects the health data, connects and offloads to the edge node at doctor side. The edge device accepts the connection request and performs the critical-level assessment based on the received data as well as personal information, historical medical data and other parameters. The patient receives the risk condition analysis feedback from the edge node directly to his/her device. Under normal conditions, the edge nodes send back the computed results directly to the patient IoT device. If critical, the edge nodes will allocate the resources to the nearest medical computational servers (MCSs). The doctor edge device accepts the connection request and carries out an extensive and accurate medical pinpointing and analysis of the results. The doctor sends the prescription and requisites from his/her device to the patient's IoT device after carrying out extensive and accurate medical pinpointing and analysis of results. Figure 2 depicts the flow sequence of activities between all the entities of the system.

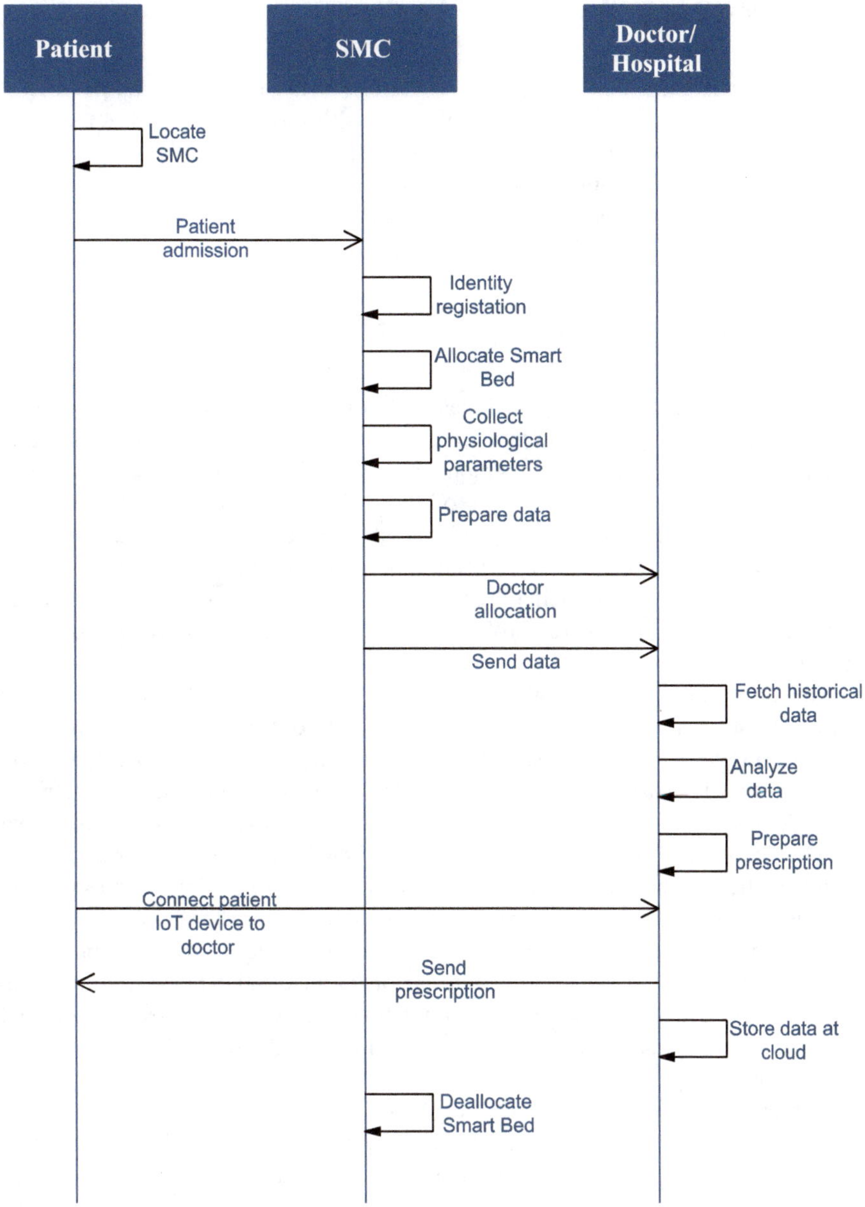

Fig. 2 Workflow of patient visit to SMC

4 Proposed Scheme for SMC Prediction

The raw data is gathered from specific zones of Ahmedabad city and processed in a vector matrix. The data set consists of potential locations identified in the study area and divided into four geographical regions of interest viz. north, south, east and west. A detailed description of the implemented genetic algorithm for prediction of location of SMC is discussed in the below sections.

4.1 Genetic Algorithm

Genetic algorithm (GA) is a special class of evolutionary computation algorithms. Genetic algorithm, a guided random search technique, is a meta-heuristic algorithm that utilizes the concept of population. It mimics the mechanism of Darwinian theorem which states that the chance of survival of the organism depends on its fittest values. If the fitness value is higher, the chances of survival are more in an evolution process. GA has been used traditionally to find solution to problems like travelling salesman problem (TSP). GA uses bottom-top approach and provides solution to complex and search problems of space having extremely high possible combinations.

The above-mentioned algorithms are developed on a random process and produce a set of multiple solutions in place of only one solution to the problems involving rare information. In aspects of GA, this is called as population. Population is made up of sequences known as chromosome or individual; each item in the chromosome is known as gene. This set of chromosomes creates an initial population of the GA program and tested for quality. A metric called fitness function evaluates the best possible solution for individual in the population and assigned a fitness value. A high-value chromosome is selected from the pool of available chromosomes based on this fitness function. These selected chromosomes are further processed for crossover and mutation operations, and the best chromosomes are chosen in order to create a new population. The whole process is executed repeatedly during the operation of the algorithm until the desired and best result is achieved.

Our proposed methodology uses evolutionary GA approach and geographical information system (GIS) for solving optimal location problem. The best site/location search is required for the efficient urban planning. The location planning involves n sites accommodating n facilities (e.g. school, medical centres etc.) in $N \times N$ cell space. For example, to accommodate 10 facilities in 100×100 cell space will require $\frac{10,000!}{10! \times 9990!} = 2.74 \times 10^{33}$ possible combinations. This is just a simple problem that accounts for enormous computation time and resources. The above can be intelligently solved using GA. The proposed method also exploits GIS to retrieve spatial information. GA employs methods like chromosome selection, crossover, mutation and elitism. The main stages employing optimal procedures for establishing n SMC based on GA and GIS processes are depicted in Fig. 3.

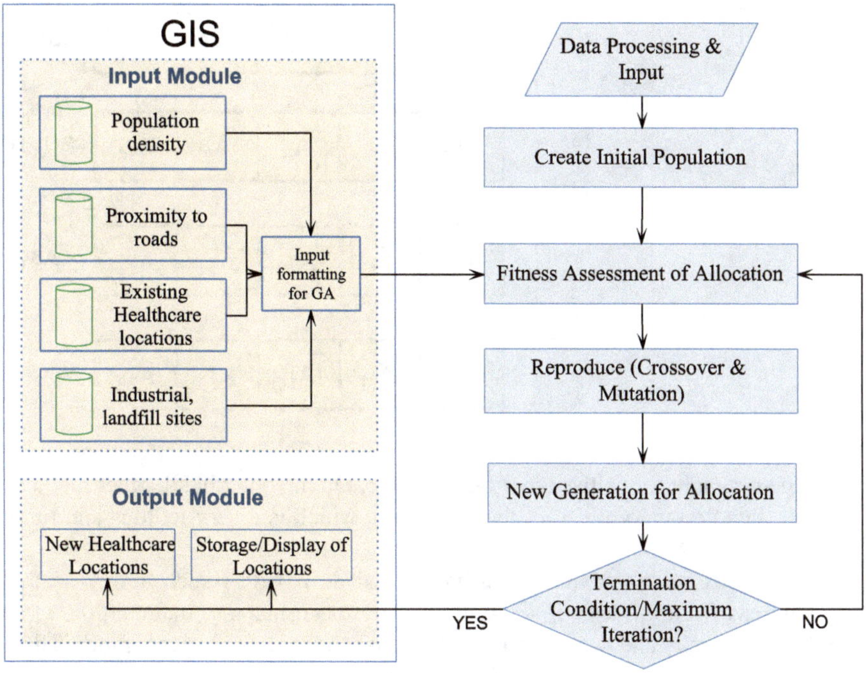

Fig. 3 Genetic algorithm with GIS for facility allocation

4.2 *Population Initialization*

In GA, the basic component is an individual or chromosome which is represented as an array to define variables of a search problem for finding optimum solution. GA depends on the generation of the population until a certain decision threshold is reached. Here, chromosome $C = [A_i; i = 1, 2, \ldots, 20]$ represents sites for implementing SMC from different zones of Ahmedabad city as explained previously and shown in Fig. 4 as a sample. Random start method is used which will provide initial population for the GA. The selection for the size of population is trade-off between finding of best solution and computational time.

4.3 *Selection*

After each generation, some of the selected chromosomes are used for generation of new population. In each generation, some part of the existing population is chosen to develop upcoming generation. The chromosomes with the best fitness are selected to improve the population and have a higher chance to go into the next generation. Certain selection methods check the fitness value of each possible solution and

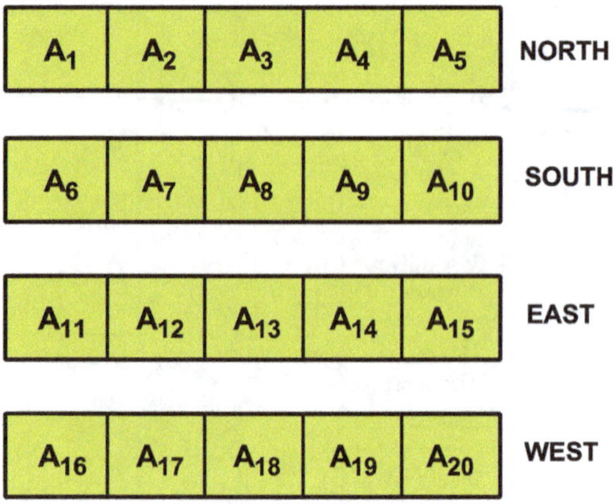

Fig. 4 A sample chromosome

select the best solution with higher fitness value. On the other hand, other methods calculate sample randomly from the population which is a very tedious and time-consuming process.

Many functions are stochastic in nature and designed in such a way so that few solutions from the less fit solutions are chosen. It helps to maintain the large population diversity and avoid stuck into the local optima for less fit solutions. There are two main methods that are frequently used and well defined in the literature viz. roulette wheel selection and tournament selection.

Roulette Wheel Selection Method Roulette wheel is a stochastic selection method. In this technique, the chromosomes present in the population are allocated specific position on the roulette wheel based on their fitness value. Each chromosome is provided a sector of wheel, where sector represents the fitness of the chromosome. In order to obtain N individuals in the mating pool, the wheel is spun N times until the required number of chromosomes is chosen. The probability of selection of chromosome is completely determined by its fitness value; chromosome with higher fitness value has higher probability of going into the next generation. Figure 5 shows the roulette wheel selection procedure.

In roulette wheel selection, the average fitness of the population for ith generation is determined as

$$Fr_{i,j} = \frac{\sum_{j=1}^{n} Fr_j}{N} \tag{1}$$

where N denotes the number of generations and Fr_j is the fitness of individual j. Therefore, the probability for selecting the jth individual is

$$Pr_j = \frac{Fr_j}{\sum_{j=1}^{N} Fr_j} \tag{2}$$

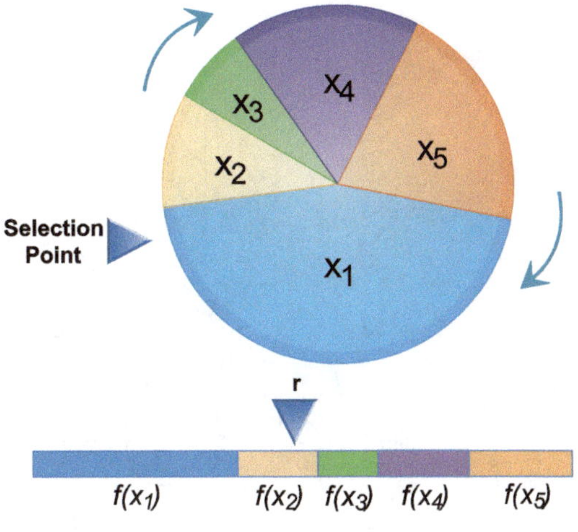

Fig. 5 Roulette wheel selection

Tournament Selection Method The rate of convergence in GA is mainly calculated by the selection pressure. If the selection pressure (a probabilistic measure of a chromosome's likelihood of participation) is very low, the rate of convergence will be slow, and the time taken by GA to determine the optimal solution is very large. If the selection pressure is very high, GA may prematurely converge to an incorrect (sub-optimal) solution.

Tournament selection includes execution of *s* tournaments from which *s* few chromosomes are selected randomly from the population. Among *s* tournament competitors, the chromosome with a higher fitness value is declared as the winner of the tournament. Selected winners are added into the mating pool. Now, mating pool includes all tournament winners with higher average fitness compared to the average fitness of the population. Selection pressure is determined by fitness difference, which helps GA to enhance the fitness of upcoming generations. Selection pressure can be modified by updating the size *s* of tournament. Figure 6 shows the tournament selection methodology.

In our research, the method of roulette wheel selection is used that provides the possibility to select each chromosome in the population.

4.4 Objective Functions

Real-life optimization problems are associated with multiple objectives that are used to find solutions where the results of designed objective functions provide an optimum value close to the real solutions. Multiple objectives conflict with each other but provide superior optimization results compared to a single objective.

SMC establishment aims to provide low-cost hospital alike facility to patients while engaging low cost, less human intervention and enabling real-time analysis

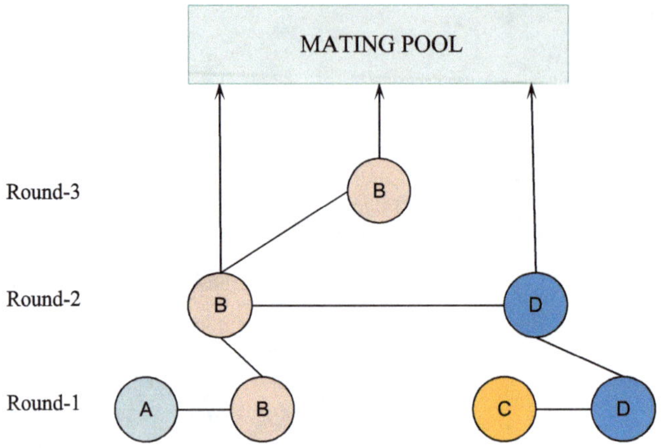

Fig. 6 Tournament selection using mating pool

and diagnosis of disease, reducing the need for physician visits. The fitness function is one of the core criteria of GA because it determines the chance of selection. Through fitness function, fitness value (a number) is calculated for each chromosome present in the population and assigned to the latter. Fitness functions are designed on the basis of fulfilment of criteria to the planning objectives and selection criteria. Four objectives have been identified for SMC location modelling and described as below.

Maximize the Accessibility for Population A fitness function is used to determine the total population that can be facilitated by m facilities. It is always required to serve maximum population as possible through these m facilities. It is similar to identifying best places to target m bullets to maximize death toll. The objective (or fitness) function of maximizing accessibility can be written below as

$$f_1 = \sum_{i=1}^{m} \sum_{x_i-\left(\frac{l-1}{2}\right)}^{x_i+\left(\frac{l-1}{2}\right)} \sum_{y_i-\left(\frac{l-1}{2}\right)}^{y_i+\left(\frac{l-1}{2}\right)} P_{den}(x, y) \cdot A_0 \cdot f^{'}(dis) \tag{3}$$

where x_i and y_i represent ith site location coordinates, m represents the allocation of total sites of the area, l denotes the neighbourhood window for adding the whole population to facilitate, $P_{den}(x, y)$ denotes the population density which is varying dynamically, A_0 represents individual cell area and $f^{'}(dis)$ represents the density decay function. We have considered Gaussian-based decay function [29] for our research, which is given as $f^{'}(dis) = e^{-k\sqrt{(x-x_i)^2+(y-y_i)^2}}$ in terms of x_i and y_i, and k is a decay function coefficient whose value ranges from 0.5 to 1.2. The rate of accessibility will decrease exponentially from facility site. The varying population density is determined by negating the served population from the current population.

In our research, we assume that the demand at each site does not exceed facility capacity and the demand is allocated to at most one facility.

Minimize the Distance of Site to Roads The objective (or fitness) function to minimize distances between selected medical centres (sites) to the closest roads can be written below as

$$f_2 = \sum_{i=1}^{n} dis_i \tag{4}$$

where n denotes the selected sites, and dis_i denotes the distance between selected site and closest road. The function f_2 is minimized by fulfilling the below distance criteria:

$$d_{ij} > D_{min} \tag{5}$$

where D_{min} denotes the distance between possible combinations of chosen sites, and $d_{ij} = \sqrt{(x_i - x_j)^2 + (y_i - y_j)^2}$ calculates the distance between the ith and jth possible locations (sites).

Minimize the Transportation Cost The transportation cost is obtained by totaling the distance between the medical site facility and demand cell/points, factored by weighing points given by the corresponding population. To minimize the energy consumption for transportation is the objective of this function. It can be expressed below as

$$f_3 = \lambda \cdot \sum_{i=1}^{n} \sum_{j=1}^{n} d_{ij} \cdot p_i \cdot Z_{ij} \tag{6}$$

where p_i is the population in point i, and λ is a scaling constant. Z_{ij} is '1', if clinic at location j provides service to the patients at point i and '0' if not.

Minimize the Location Criteria The spatial analysis is conducted to determine SMC criteria; evaluation of fitness for the study area is based on the criteria and factors set previously. Each chromosome is compared with the given location criteria for SMC establishment formulated as below:

1. No SMC within 1.5 km vicinity of the already existing healthcare facility
2. No SMC within 500 m of industrial areas
3. No SMC within 500 m of landfill sites

The objective function can be determined by the following equation:

$$f_4 = \frac{1}{n} \sum_{i=1}^{n} a_i, \quad \forall n \in C \tag{7}$$

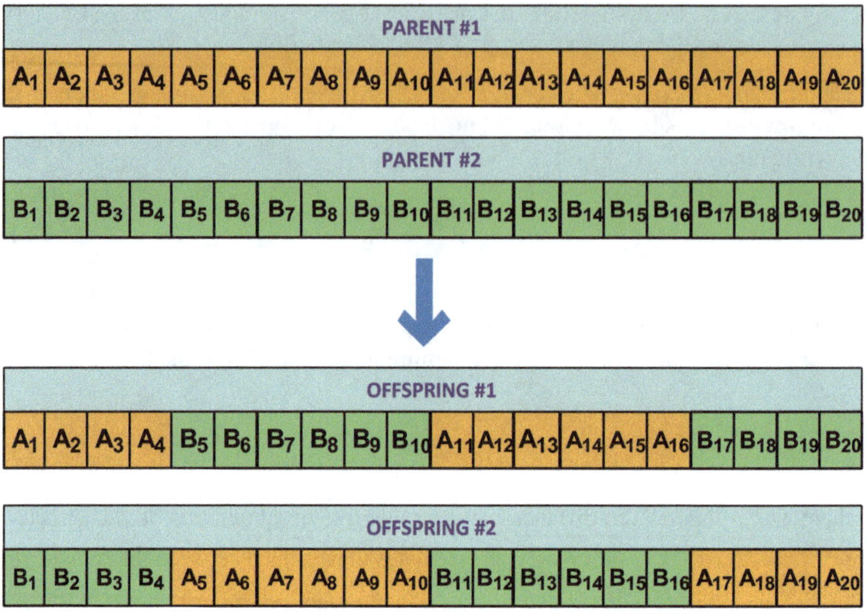

Fig. 7 Crossover

where n is the number of chromosome, specifying the criteria $C = (0, 1)$, and a represents the sum of conditional values.

4.5 Crossover and Mutation

Crossover is a genetic operation that takes parent chromosomes to generate a new offspring from one generation to the next. Random genes are taken from the mating chromosomes, which results into offspring with superior characteristics. Multi-point (two-point) crossover is selected for our approach. The fittest location among the two parents is selected as two points for crossover to produce the new offsprings. Figure 7 shows the parents and obtained child chromosomes.

In mutation, random changes in the genetic material of the chromosomes are required to avoid stagnation at local optima and minima. To diversify the chromosome from its basic level, mutation operator is used. In our research, classical mutation operator is applied to the solutions, generated from crossover operation. A new chromosome replaces the repeated chromosome with minimum value by random selection having the maximum fitness value. The process is repeated until a final normalized value is obtained in all the previous iterations. As shown in Fig. 8, the substitution method is used for our approach where area A_1 is replaced with A_5.

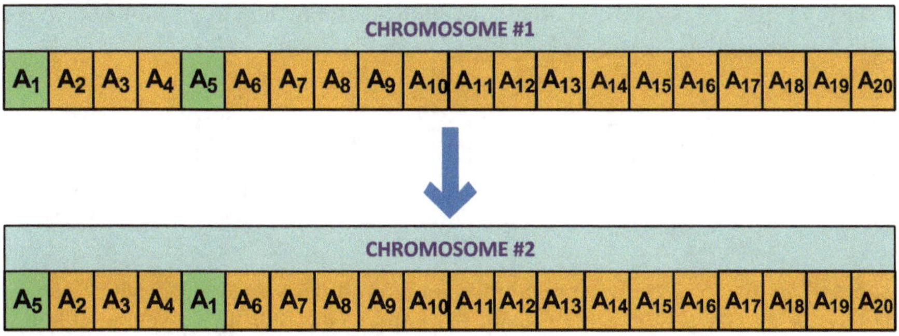

Fig. 8 Mutation

4.6 Multi-Criteria Objective Model for SMC Prediction

A number of research works have developed and demonstrated the methodologies to compute multiple objective problems in GIS. Multi-criteria evaluation (MCE) technique deals with multi-objective problems to achieve optimization of multiple objectives at the same time and provide decision-making process. The problems of optimum medical facility establishment using GA are subjected to multi-objective functions through multiple constraints for the evolutionary process. The conflict and influence of each function may decrease the performance improvement of a target and impact other target performance. However, single-objective function can be designed from the above-mentioned fact using a combination of weighted coefficients. It is impossible to achieve multi-criteria objective by integrating GIS and MCE due to enormous computational requirements. The integration of MCE with GA is the need for solving problems under the afore-mentioned circumstances. The MCE model transformed into a single objective can be expressed as follows:

$$
\begin{aligned}
f_5 &= \sum_{i=1}^{4} w_i f_i, \quad \left\{ w_i = [0,1], \sum_i w_i = 1 \right\} \\
&= w_1 f_1 + w_2 f_2 + w_3 f_3 + w_4 f_4
\end{aligned}
\tag{8}
$$

where w_i is the weight of an individual objective function. The score value of individual functions must be normalized before the actual calculation. The normalization can be performed by varying the scaling of the function between the range of [0,1] by minimum and maximum values through a linear derivative transformation as below:

$$
f_i' = \frac{f_i - f_{imin}}{f_{imax} - f_{imin} + 1} \tag{9}
$$

Using Eq. 9, the normalized fitness function corresponding to the multi-objective can be expressed by the weight coefficient method as below:

$$F_5 = \sum_{i=1}^{4} w_i F_{i(n)}, \quad \left\{ w_i = [0, 1], \sum_i w_i = 1 \right\} \tag{10}$$
$$where \quad F_i = f_i / f_i^{'}$$

where f_{imax} and f_{imin} are the maximum and minimum values of f_i, which can be derived using GA program inclusively from the single-objective problem solution using Eqs. 3, 6, 4 and 7. The SMC locations are selected based on the roulette wheel method to form initial chromosome population. The probability of selection is proportional to the normalized fitness value; a higher value is assigned the larger slot size. The process is repeated until all points are assigned to select next parent.

The new generation follows the similar procedure as their parents follow to achieve the best population position. An addition strategy is employed to generate the new population for every iteration. In this strategy, to keep best quality of the past generation, elite offsprings are inserted into the current population. The best fitness value increases rapidly along with time, which results in better offsprings in the evolutionary process. The offsprings are ordered based on the fitness values, and the process is repeated until the average fitness in each generation less than a particular threshold or iterates until the maximum number of iterations is attained. The program terminates when $F_5(t+1) - F_5(t) < \epsilon$, where ϵ is a small value. The best elite chromosome during the final search will provide the optimal allocations of the m sites of SMC. The $\{x, y\}$ coordinates for the optimally selected sites can be expressed by the chromosome as $CM^{opt} = [x_1^o y_1^o, x_2^o y_2^o, \ldots x_m^o y_m^o]$.

The evolutionary process is objective-oriented as the fitness functions reflect competitive objectives. Optimization allows the usage of simple mathematical problem formulation and makes the calculation simpler and advantageous over conventional methods (1-median, p-median, set covering problem etc.). It provides feasibility for inputting geographical information into the GA program.

5 Analysis and Experimental Results

As per the Organization for Economic Co-operation and Development (OECD) data available for 2017, India reportedly has only 0.53 beds available per 1000 people. Especially in the Gujarat state, with over 6.82 crore population (as of 2018), it has only 16,375 beds. Ahmedabad is the 1st largest city in Gujarat with the population of around 92.5 lakhs (2019). Ahmedabad city stands third highest in the number of cases due to recent COVID-19 outbreak and has been identified as one of the major hotspots. It becomes difficult to increase the testing/care facilities due to large population densities in cosmopolitan cities especially covering the slum pockets. In critical conditions, a patient can visit to the nearest facility and utilizes

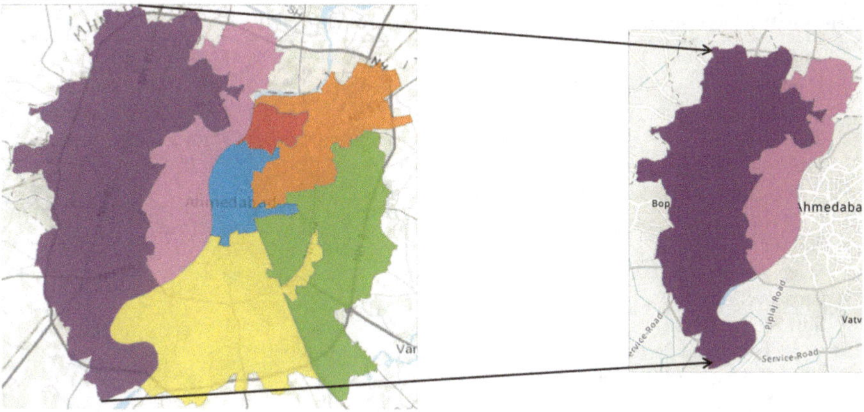

Fig. 9 Study area

the healthcare facility. The transportation facility can be arranged by the patient by himself/herself or can utilize the ambulance facility provided by the healthcare authorities. Partitioning of city into various regions is necessary to avoid huge inclusion of patients from one micro-region to another and increase quality thereby reducing the cost.

For smart cities, consideration of problem of finding healthcare facility is necessary in a complex MO (multi-objective) problem rather than a single-objective problem. The complexity of identifying the location increases due to inclusion of real-world constraints such as environmental, economic, traffic, density of area and various other factors. The objective of this work is implemented for the specific zones of Ahmedabad district, which lies between $23°02'48.28''N$ and $72°37'08.06''E$. Figure 9 shows the study area.

5.1 Analysis

Algorithm 1 depicts the GA steps under consideration. Initial population is taken randomly as input. The fitness value is calculated for each chromosome using multi-objective fitness function. Parents $p1$ and $p2$ are selected from the population to produce offsprings. Multi-point crossover is performed to generate variability in chromosome. $p1$ and $p2$ are inputs to the crossover function $multipointcrossover$ that provides a new set of population. The population generated after crossover operation $(Popcrossover)$ is mutated to increase the diversity. The chromosomes generated after mutation operation $mutation$ are checked for feasibility to replace the old ones with a new set of chromosomes. The former will be evaluated using the fitness function, and the best fit chromosomes will be returned as an optimal solution for the required number of generations. The time complexity of the algorithm is $O(n \log m)$.

Algorithm 1 Genetic algorithm for establishment of SMC

```
Input: Population                                ▷ List of chromosomes as candidate solution
Output: Best chromosome
initialize population randomly
int Pop[m][n]              ▷ m,n denotes size of population and chromosome, respectively
for j ← 1 to size(Pop) do
    calculate fitness value of each chromosome
    fitness[j] ← fitnesscal(Pop)
end for
Select m best chromosome on the basis of fitness
Crossover operation
for i ← 1 to m do
    Parent-1 and Parent-2 select randomly from the population
    p1 ← rand(Pop[m][n])
    p2 ← rand(Pop[m][n])
    Popcrossover ← multipointcrossover(p1, p2)    ▷ generated child using multi-point crossover
end for
updatepop ← Popcrossover
Mutation operation
for k ← 1 to n do
    Select chromosome from updated list
    newchromosome ← mutation(chromosome[k])
    if newchromosome is feasible then
        replace with best chromosome
    end if
end for
Pop_best contains m best chromosomes
Select global best from the available best chromosome
for l ← 1 to m do
    if fitness(l) > globalbest then
        globalbest ← fitness(l)
    end if
end for
return globalbest
```

5.2 *Results*

The study area selected for the establishment of smart medical centres (SMC) is New West zone and West zone of Ahmedabad district, Gujarat State, India as highlighted in Fig. 9 in Sect. 2 with a population of around 25 lakhs projected from 2011 population census (Ahmedabad Municipal Corporation (AMC)). The population data is available for district blocks/zones. Data sets for different features such as streets, demand points, nearest roads, industrial sites, population density, existing healthcare facility and transportation conditions were collected and considered in location establishment process.

ArcGIS is the popular GIS software for visualizing and creating spatial 2D maps, geographic data and information and managing geographic information in a database. Various ArcGIS tools were used to manage data demand points, medical centre locations, calculate distance from streets, site suitable map and population density in each district zone. The database for population density, closeness to roads and existing healthcare facilities were provided in a raster format in a cell size of $100\,m^2$ consisting of grid layer for optimization of various objective functions for each cell size and inputted to the GA modelling program.

Based on the digitized data of map database, the existing healthcare facilities were identified in the study area. These facilities were modelled in respective grid cell, and the maps of the surrounding primary and secondary streets were converted to points to determine the accessibility distance. In the study area, it was found that there are only 50 health facilities that serve an average settlement for nearby as well as far areas. Considering the events of emergency, rapid urbanization of the area as well as more reach to the patients by healthcare provider/doctor and vice versa, additional medical centres encompassing smart IoT technology were the necessity to support fast and reliable medication and diagnosis.

Each new smart medical centre (SMC) in grid cell area represents chromosomes in the population. These locations are binary encoded and represented as string as a pair of $\{x, y\}$ coordinates for the ith medical centre. The fitness function parameters include neighbourhood size (l) for adding population and coefficient of decay function (k). Equation 3 dictates the fitness function formulated to serve the total population for m facilities according to an exponential decay function. Here, l and k are set to 25 and 0.08, respectively. In GA, parameters are crucial to provide the best solution. The following GA parameters are considered.

- Size of population: 200
- Size of chromosome: 20
- Maximum generations: 200
- Rate of crossover: 0.9
- Rate of mutation: 0.01

The algorithm was executed on an Intel®Core™i7 3.2 GHz PC running Windows 10 with 4GB RAM. The genetic algorithm was developed in MATLAB®to solve the single-objective as well as multi-objective criteria for optimization. MATLAB®package provides the development toolkit to utilize necessary functions.

The input data was prepared, and the GA model was executed by applying parameter setting as mentioned above. In total, we realized fifty solutions resulting from multi-objective optimization for the establishment of SMC. A comparative study was done to visualize the effect of optimization based on evaluation of only single-objective functions. The first objective function was assigned a higher weight value, and the corresponding lower weight value was assigned to the remaining functions. The same is repeated for the remaining three objective functions.

Table 1 depicts the optimum solutions and associated fitness values for each objective function. To maximize the accessibility of population (first objective function OF_1), a higher weight ($w_1 = 0.85$) is considered. The average travelling distance is comparatively low; however, the number of SMC increases more towards the vicinity of residential streets and already existing healthcare facility resulting in increased cost and redundancy. Figure 10a shows the effect of maximizing the accessibility to the population for 20 SMCs. Similarly, when second objective

Table 1 Single-objective function optimization results

Sr. no.	Weight (w)	Sol. no.	Sites
1	[0.85, 0.1, 0.1, 0.1]	20	56, 99, 25, 3, 169, 29, 192, 106, 42, 101, 55, 6, 16, 22, 149, 75, 142, 86, 58, 189
2	[0.1, 0.85, 0.1, 0.1]	36	56, 99, 187, 169, 25, 32, 106, 42, 9, 68, 15, 101, 94, 22, 16, 75, 61, 8, 52, 18
3	[0.1, 0.1, 0.85, 0.1]	40	94, 56, 24, 30, 133, 187, 61, 13, 68, 188, 103, 101, 163, 45, 37, 22, 75, 1, 40, 84
4	[0.1, 0.1, 0.1, 0.85]	69	56, 26, 67, 41, 71, 187, 121, 68, 179, 160, 134, 45, 28, 90, 22, 73, 10, 46, 89, 70

(OF_2) with a higher weight ($w_2 = 0.85$) is considered, the analysis shows that by selecting solution no. 36, the average travelling distance is within 500 m to the nearest site; sites (101,9,18) are located near to industrial areas, which tends to be hazardous, and sites (169,75) are located in expensive areas; however, the overall solution is expensive because of the increased number of clinics. Figure 10b shows the effect of minimizing the transportation cost for 20 SMCs. For the third objective function (OF_3) with a higher weight ($w_3 = 0.85$), solution no. 40 provides an optimal solution as shown in Fig. 10c. The analysis results show that sites (22,45,188,84,103,94) are within the permissible range of non-acceptable regions (such as highways, heavy traffic borders, close roads) and having acceptable distance (D_{min} set to 1.5 km) between two successive pairs of established SMC. Similarly, the fourth objective (OF_4) with a higher weight ($w_4 = 0.85$) is considered, for which solution no. 69 provides the best solution; all the SMCs were located at an average distance of 1 km when industrial sites and waste dump yards were considered. The fourth objective function is determined by constraints through individual chromosome and determined by applying statistical mean weight that measures the fitness of each chromosome and selects the chromosome with a minimum mean value. In our research, we have selected the locations of SMC as the initial coordinates for iteration of the algorithm to minimize the above function. Figure 10d shows the optimal locations according to the fourth objective function. Table 1 shows the weight vectors (w), solution serial number (Sol. no.) and site index obtained after optimization using single-objective functions.

To compare the solutions, normalized values of all the four objective functions were obtained. Objective function values after normalization were summed together

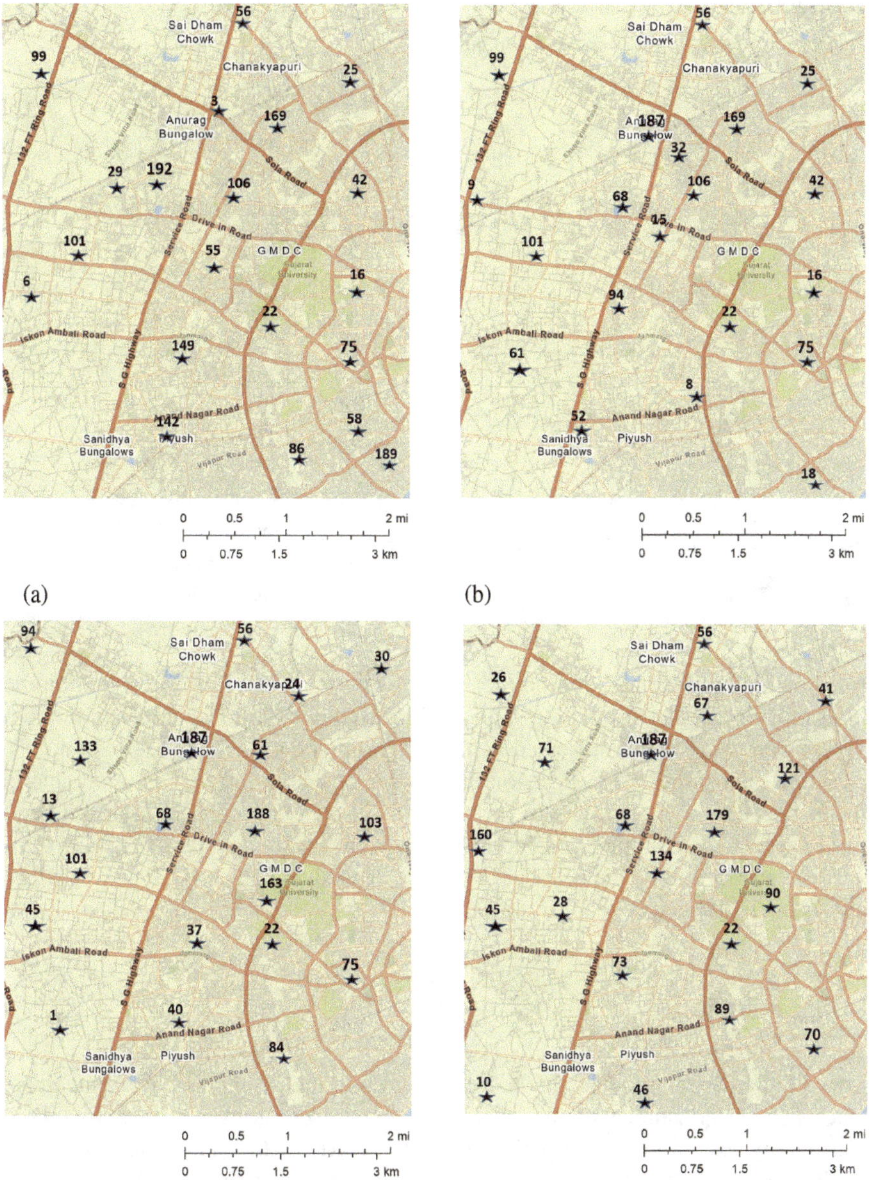

Fig. 10 Selected sites and assigned objectives based on distances. (**a**) Sites according to the first objective. (**b**) Sites according to the second objective. (**c**) Sites according to the third objective. (**d**) Sites according to the fourth objective

Table 2 Details of four solutions based on multi-objective optimization

Sr. no.	Normalized objective functions $[F_{1(n)}, F_{2(n)}, F_{3(n)}, F_{4(n)}]$	Sol. no.	Sites
1	[0.122, 0.165, 0.231, 0.141]	23	9, 4, 123, 6, 168, 49, 91, 99, 48, 174, 189, 120, 111, 172, 153, 32, 29, 50, 97
2	[0.234, 0.034, 0.139, 0.321]	102	25, 67, 36, 129, 157, 163, 19, 1, 78, 80, 20, 5, 120, 39, 99, 48, 88, 70, 46, 91
3	[0.023, 0.126, 0.176, 0.2795]	45	3, 46, 37, 124, 65, 80, 44, 99, 105, 69, 87, 120, 118, 123, 161, 194, 135, 47, 90, 78
4	[0.421, 0.261, 0.089, 0.209]	35	7, 30, 56, 108, 119, 49, 63, 89, 174, 120, 118, 140, 46, 93, 127, 66, 52, 150, 104, 97

for providing individual solution (denoted by t). Then, solutions are sorted on the basis of their calculated function value, and finally four solutions with the minimum values were chosen and presented in Table 2. From the available pool of solutions, based on their preferences, decision-makers can select global best solution among them. We can assign a weight to each objective function according to priority and importance. For example, considering the weight vector as {0.5, 0.4, 0.3, 0.2}, solution no. 45 will be the best one. The final normalized objective function values are shown in Table 2.

6 Conclusion

Location prediction models play a crucial role in establishing healthcare facilities. One of the establishments that can leverage maximization in the health treatment and quality care for modern patient–doctor scenario is the smart medical centre (SMC). In our research, we have proposed the architecture of SMC based on IoT and edge management technology. The resource distribution mechanism enabled by edge computing effectively optimizes the latency requirements for patients under different medical conditions as well as caters to remote monitoring of latter by healthcare institutions/doctors to implore appropriate actions for real-time diagnosis and medications.

Genetic algorithms (GAs) have capability of producing optimized results for optimal location search of a number of healthcare facilities under complex situations. GA effectively uses the natural biological process of evolution and tends to establish an optimal solution to a given complex problem. The proposed scheme uses multi-objective optimization approach based on GA to find the optimal trade-offs between the multiple objectives in determination of SMC. Four objective functions were defined to maximize the accessibility to the population and minimize the transportation costs, as well as to minimize the distance to the roads and avoidable/hazardous or redundant locations to identify the location and an optimum number of SMCs. The functions were maximized/minimized to study the overall effect of only single-objective functions on the remaining ones. If maximizing the accessibility and minimizing the transportation cost are considered for example, then redundancy in medical centres will increase, which can increase the overall cost.

The multi-objective criteria were evaluated using the above single-objective functions for two zonal areas (New West zone and West zone) of Ahmedabad city through the genetic algorithm integrated with GIS to obtain spatial data and find the Pareto-optimal non-dominated solutions. The spatial data is used to evaluate the fitness function to retrieve optimal solutions. The multiple optimal solutions obtained represent the best trade-offs between conflicting objectives. While comparing the multiple optimal solutions, it is required to consider various types of relative importance of each objective function. All the relative importance associated with each objective function is to be considered for comparing multiple optimal solutions, and the several solutions were obtained. Based on visualization, investigation and relative importance/preference of the optimal solutions through different locations of sites and numbers, it can help decision-makers in location planning for new medical centres. However, the objectives and corresponding functions may change based on the criteria of the particular city, when multi-objective criteria were applied to other case studies.

In upcoming days, location allocation problem of healthcare facility will be one of the most priority planning problems. Therefore, it is required to focus to construct and execute methods to support the decision-making process of designers in establishing new healthcare centres. GA-based multi-objective optimization methodology proposed in this research represents a trade-off among different conflicting objectives and further can be used with objectives in the city under interest.

The proposed methodology can be combined with other optimization techniques such as Ant Colony Optimization (ACO) and Particle Swarm Optimization (PSO) to achieve more objectives and improved performance. The adaptability in the methodology can be included by considering dynamics of various constraints included in the objective functions as well as variability in the chromosome size to provide highly efficient and cost-effective solutions for establishment of SMC.

References

1. Akmandor, A.O., Jha, N.K.: Smart health care: An edge-side computing perspective. IEEE Consumer Electronics Magazine **7**(1), 29–37 (2018). https://doi.org/10.1109/MCE.2017.2746096
2. Aloi, G., Caliciuri, G., Fortino, G., Gravina, R., Pace, P., Russo, W., Savaglio, C.: A mobile multi-technology gateway to enable IoT interoperability. In: 2016 IEEE First International Conference on Internet-of-Things Design and Implementation (IoTDI). pp. 259–264 (2016). https://doi.org/10.1109/IoTDI.2015.29
3. Azrour, M., Mabrouki, J., Chaganti, R.: New efficient and secured authentication protocol for remote healthcare systems in cloud-iot. Security and Communication Networks **2021**, 1–12 (2021). https://doi.org/10.1155/2021/5546334
4. Azrour, M., Mabrouki, J., Guezzaz, A., Farhaoui, Y.: New enhanced authentication protocol for internet of things. Big Data Mining and Analytics **4**(1), 1–9 (2021). https://doi.org/10.26599/BDMA.2020.9020010
5. Beheshtifar, S., Alimohammadi, A.: Multi-objective evolutionary algorithm for modeling of site suitability for health-care facilities. Health Science Journal **7**, 209–217 (01 2013)
6. Burkey, M., Bhadury, J., Eiselt, H.: A location-based comparison of health care services in four U.S. states with efficiency and equity. Socio-Economic Planning Sciences **46**(2), 157–163 (2012). https://doi.org/10.1016/j.seps.2012.01.002
7. Chen, M., Li, W., Hao, Y., Qian, Y., Humar, I.: Edge cognitive computing based smart healthcare system. Future Generation Computer Systems **86**, 403–411 (2018). https://doi.org/10.1016/j.future.2018.03.054
8. Daskin, M.S., Dean, L.K.: Location of Health Care Facilities, pp. 43–76. Springer US, Boston, MA (2004), https://doi.org/10.1007/1-4020-8066-2_3
9. Demir, E.: Approach for siting a support facility for transporting supplies in emergency cases in the republic of Bulgaria. MATEC Web Conf. **234**, 06001 (2018). https://doi.org/10.1051/matecconf/201823406001
10. Demir, E., Kockal, N.: Iterative methodology on locating a cement plant. Journal of Inequalities and Applications **2019** (06 2019). https://doi.org/10.1186/s13660-019-2122-7
11. Gómez, J., Oviedo, B., Zhuma, E.: Patient monitoring system based on internet of things. Procedia Computer Science **83**, 90–97 (2016). https://doi.org/10.1016/j.procs.2016.04.103
12. Greco, L., Percannella, G., Ritrovato, P., Tortorella, F., Vento, M.: Trends in IoT based solutions for health care: Moving AI to the edge. Pattern Recognition Letters **135**, 346–353 (2020)
13. Greco, L., Ritrovato, P., Xhafa, F.: An edge-stream computing infrastructure for real-time analysis of wearable sensors data. Future Generation Computer Systems **93**, 515–528 (2019). https://doi.org/10.1016/j.future.2018.10.058
14. Hakimi, S.L.: Optimum locations of switching centers and the absolute centers and medians of a graph. Operations Research **12**(3), 450–459 (1964). https://doi.org/10.1287/opre.12.3.450
15. Hakimi, S.L.: Optimum distribution of switching centers in a communication network and some related graph theoretic problems. Operations Research **13**(3), 462–475 (1965). https://doi.org/10.1287/opre.13.3.462
16. Hegde, C., Jiang, Z., Suresha, P.B., Zelko, J., Seyedi, S., Smith, M.A., Wright, D.W., Kamaleswaran, R., Reyna, M.A., Clifford, G.D.: Autotriage - an open source edge computing raspberry pi-based clinical screening system. medRxiv (2020). https://doi.org/10.1101/2020.04.09.20059840
17. Islam, S.M.R., Kwak, D., Kabir, M.H., Hossain, M., Kwak, K.: The internet of things for health care: A comprehensive survey. IEEE Access **3**, 678–708 (2015). https://doi.org/10.1109/ACCESS.2015.2437951
18. Kaur, A., Jasuja, A.: Health monitoring based on IoT using raspberry pi. In: 2017 International Conference on Computing, Communication and Automation (ICCCA). pp. 1335–1340 (2017). https://doi.org/10.1109/CCAA.2017.8230004

19. Li, X., Yeh, A.G.: Integration of genetic algorithms and GIS for optimal location search. International Journal of Geographical Information Science **19**(5), 581–601 (2005). https://doi.org/10.1080/13658810500032388
20. Mabrouki, J., Azrour, M., Driss, D., Farhaoui, Y., Souad El, H.: Iot-based data logger for weather monitoring using Arduino-based wireless sensor networks with remote graphical application and alerts. Big Data Mining and Analytics **4**(1), 25–32 (2021). https://doi.org/10.26599/BDMA.2020.9020018
21. Mabrouki, J., Azrour, M., Farhaoui, Y., El Hajjaji, S.: Intelligent system for monitoring and detecting water quality. In: Farhaoui, Y. (ed.) Big Data and Networks Technologies. pp. 172–182. Springer International Publishing, Cham (2020)
22. Mabrouki, J., Azrour, M., Ghizlane, F., Driss, D., Souad El, H.: Intelligent monitoring system for biogas detection based on the internet of things: Mohammedia, morocco city landfill case. Big Data Mining and Analytics **4**(1), 10–17 (2021). https://doi.org/10.26599/BDMA.2020.9020017
23. Neema, M., Ohgai, A.: Multi-objective location modeling of urban parks and open spaces: Continuous optimization. Computers, Environment and Urban Systems **34**(5), 359–376 (2010). https://doi.org/10.1016/j.compenvurbsys.2010.03.001
24. Osinuga, I.A., Oyebade, V.O., Otun, W.O.a.: Integrated technique for the optimization of healthcare facility problem. International Journal of Applied Operational Research **10**(1) (2020), http://ijorlu.liau.ac.ir/article-1-606-en.html
25. Peng, Q., Afshari, H.: Challenges and solutions for location of healthcare facilities. Industrial Engineering & Management **03** (01 2014). https://doi.org/10.4172/2169-0316.1000127
26. Rao, P.K., Omar, A.: Dedicated real time monitoring system for healthcare using Zigbee. Healthcare Technology Letters **4** (06 2017). https://doi.org/10.1049/htl.2017.0030
27. Siam, A., Abouelazm, A., El-Bahnasawy, N., El Banby, G., Abd El-Samie, F.: Smart health monitoring system based on IoT and cloud computing (12 2019)
28. Uddin, M.S., Alam, J.B., Banu, S.: Real time patient monitoring system based on Internet of Things. In: 2017 4th International Conference on Advances in Electrical Engineering (ICAEE). pp. 516–521 (2017). https://doi.org/10.1109/ICAEE.2017.8255410
29. Wang, F., Tang, Q.: Planning toward equal accessibility to services: A quadratic programming approach. Environment and Planning B: Planning and Design **40**(2), 195–212 (2013). https://doi.org/10.1068/b37096
30. Xiao, N., Bennett, D.A., Armstrong, M.P.: Using evolutionary algorithms to generate alternatives for multiobjective site-search problems. Environment and Planning A: Economy and Space **34**(4), 639–656 (2002). https://doi.org/10.1068/a34109
31. Özceylan, E., Uslu, A., Erbas, M., Çetinkaya, C., İşleyen, S.: Optimizing the location-allocation problem of pharmacy warehouses: A case study in Gaziantep. An International Journal of Optimization and Control: Theories & Applications (IJOCTA) **7**, 117 (01 2017). https://doi.org/10.11121/ijocta.01.2017.00373

Index

A

Accuracy (ACC), 135
Activated sludge process, 20
Actuator module, 109
Adsorption, 29
Agricultural automation technology, 104
Agricultural production, 104
  employment of sensors, 104
  IoT-based technologies
    actuator module, 109
    agricultural soil parameters, measurement of, 111, 113
    communication module, 109
    of designed system, 106
    light sensor, 107–108
    processing module, 106–107
    programmed algorithm, 111–112
    proposed system, 109–110
    soil moisture sensor, 108
    studies, 104–105
    system implementation, 110–111
    temperature and humidity sensors, 107
    water flow sensor, 108
Agriculture impacts of, 103
Anomaly network intrusion detection system, IoT
  computer security, 130
  data collection process, 134
  experimental study and assessments, 134–136
  hybrid approach, 133–134
  intrusion detection, 129, 130, 136
  machine learning techniques, 131–133
  security, 131
  vulnerabilities, 130
Ant Colony Optimization (ACO), 179
Aquaculture monitoring systems, 12–14
ArcGIS, 174–175
Arduino board, 110
ARM/Linux microcontroller, 10
Artichoke waste, 30
Artificial intelligence (AI), 156
Autonomous sensors, 112
AutoTriage system, 160

B

Bandwidth/capacity, 3
Big data analytics, 85, 86
Biochemical oxygen demand (BOD), 122

C

Calibration of a model, 23, 24
Centered composite designs, 71
  FA-T regeneration cycles, 75, 76, 82
  geometric location, 72
  method of computation, 73–74
  model validation, 74–79
  prediction profiler, 79–82
  regeneration cycles, 83
Colorimetric method, 105
Communication module, 109
Computer security, 130
Cover water tension index (CWSI), 140, 148
Crop irrigation technology, 146

M. Azrour et al. (eds.), *IoT and Smart Devices for Sustainable Environment*, EAI/Springer Innovations in Communication and Computing, https://doi.org/10.1007/978-3-030-90083-0

Crop water stress indicator measurement, 148
Crossover, smart medical centres, 170–171
Customer relation management (CRM), 86

D
Data gathering, 88
Data mining, 85
  for predictive analytics, 86–87
Defense Advanced Research Project Agency (DARPA), 134–135
Denial-of-Service (DoS) attack, 160
Detection rate (DR), 135
Digitized aerial image, 139
Dissolved oxygen (DO), 123
Drinking water distribution networks, 10–12

E
Elbow method, 97
Electrical conductivity, 47
Electronic thermal conductivity, 47–48
Emergency medical services (EMS), 162
Energy band structure, 46–47
Energy consumption, 3
Evolutionary algorithm, smart healthcare using, 155–156
  analysis, 173
  ArcGIS, 174–175
  contribution, 157
  genetic algorithms, 178–179
  location prediction models, 178
  motivation, 156–157
  multi-objective optimization, 178
  organization, 158
  scenario, 161–162
  SMC prediction (*see* Smart medical centres (SMCs))
  study area, 173
  workflow, 162–163

F
Figure of merit (ZT), 44
F-Score, 136
Full potential linear augmented plane wave (FPLAPW) method, 45

G
Genetic algorithm (GA), 157, 159, 178–179
  smart medical centres, 164–165
Geographical information system (GIS), 157–159, 164–165
GitHub platform, 92
Global Positioning System (GPS) technologies, 141
Google Cloud database, 105
GPS tool, 87
GPS-X software, 21, 22
Ground control station (GCS), 142
Groundwater quality, simulation of
  Western Rif of Morocco, limestone chain of, 117–119
    coefficients of biochemistry action, 121–122
    concentrations, simulation and monitoring of, 122–123
    materials and methods, 119–120
    simulation, 125
    stability analysis, 123–125
    study area, 119
    water quality status, 120–121

H
Health-related issues, 156
High-altitude platforms (HAPs), 140
Humidity sensors, 107

I
Infrared thermal sensors, 144–145
Intelligent farming system, 105
Interfacing sensors, 105
International Telecommunication Union, 103
Internet of Medical Things (IoMT), 155–156
Internet of Things (IoT), 103, 159
  agricultural production
    actuator module, 109
    agricultural soil parameters, measurement of, 111, 113
    communication module, 109
    of designed system, 106
    light sensor, 107–108
    processing module, 106–107
    programmed algorithm, 111–112
    proposed system, 109–110
    soil moisture sensor, 108
    studies, 104–105
    system implementation, 110–111
    temperature and humidity sensors, 107
    water flow sensor, 108
  anomaly network intrusion detection system for
    computer security, 130
    data collection process, 134

experimental study and assessments, 134–136
hybrid approach, 133–134
intrusion detection, 129, 130, 136
machine learning techniques, 131–133
security, 131
vulnerabilities, 130
potential environmental developments, 1
for real-time disease prediction
clusters definition and distribution, 93–94
K-means scatter plot, 94–95
performance and efficiency, 93

K
k-means algorithm, 93
K-means scatter plot, 94–95
K-nearest neighbors (K-NN), 132
Knowledge Discovery and Data Mining (KDD), 135

L
Laser scanners, 143
LiDAR, 143
Light sensor, 107–108
Light-dependent resistor (LDR), 105
Light-emitting diodes (LED), 105

M
Machine learning, 85
Machine Learning Library (MLlib), 92
Machine learning techniques, unmanned aerial vehicle, 146
Machine learning/deep learning (ML/DL), 156
MATLAB®, 175
Medical computational servers (MCSs), 162
Methyl orange (MO), 54, 60
Methylene blue (MB), 54, 60
Miscarriage prediction
risk factors, 91
state of the art, 90–91
system dataset, 91–93
Multi-criteria evaluation (MCE)
for smart medical centres, 171–172
Multilayer perceptron method, 105
Multi-objective optimization (MOO), 157
Multi-point (two-point) crossover, 170
Multispectral imageries (MSI), 140
Multispectral imaging sensors, 144
Mutation, smart medical centres, 170–171

N
Naïve Bayes algorithm, 129, 130, 132–133
Natural water resources monitoring systems, 5–7
Nitrogen–phosphorus–potassium (NPK) sensor, 105

P
Pareto analysis, 35–38
Particle Swarm Optimization (PSO), 179
Pharmaceuticals, 29
Pharmaceuticals active compounds (PhACs) removal, 29
Photo sensors, 108
Photoelectric devices, 108
Photogrammetry techniques, 145
Piroxicam
adsorption efficiency, 31
chemical structure of, 31
graphical analysis, 37
physical and chemical properties, 31
removal rate, 37–38
Precision agriculture, 139–140
thermal-based crop stress sensing, 146–148
unmanned aerial vehicle, remote sensing using, 140–141
advantages, 142–143
data collection using sensors, 142
ground control station, 142
surveillance of crops, 141
UAV control system, 142
using unmanned aerial vehicle, 143
data acquisition, 143–145
machine learning techniques, 146
photogrammetry techniques, 145
UAS data processing, 145
vegetation indices calculation, 145
Precision irrigation, 139
Prediction system, 87

Q
Quality of service (QoS), 156
Quality2E, 118

R
Raspberry Pi, 89, 105
Raspberry-Pi, 160

Reality mining, 85
  gathering process, 88, 89
  IoT implementation
    prediction process, 90
    sensors and tools, 88–89
    system implementation, 89
  for real-time disease prediction
    clusters definition and distribution, 93–94
    K-means scatter plot, 94–95
    performance and efficiency, 93
Real-time disease prediction
  evaluation and validation
    clustering validation metrics, 97–99
    using random K, 96
  IoT and reality mining for
    clusters definition and distribution, 93–94
    K-means scatter plot, 94–95
    performance and efficiency, 93
Remote sensing techniques, 141
  using unmanned aerial vehicle, 140–141
    advantages, 142–143
    data collection using sensors, 142
    ground control station, 142
    surveillance of crops, 141
    UAV control system, 142
Response surface methodology (RSM)
  adsorbates, 54–55
  adsorbents, 54
  ANOVA, 58, 59, 61
  coefficient of determination, 57–59
  design of experiments, 55
  experimental design matrix, 55, 58–60
  interaction profiles, 58, 62, 63
  optimization, 58
  prediction profiler fore, 62, 64
  quadratic polynomial model, 61
  regression model equation, 60, 62
  selected parameters, 56–57
  three-dimensional response surface, 62, 64–66
Robotic tools, 88
Root mean square error (RMSE), 123
Roulette wheel selection method, 166–167

S

Satellite multispectral imageries, 140
Sensors, 88
Silhouette method, 98–99
Simulation models, 118
Smart healthcare, evolutionary algorithm, 155–156
  analysis, 173
  ArcGIS, 174–175
  contribution, 157
  genetic algorithms, 178–179
  location prediction models, 178
  motivation, 156–157
  multi-objective optimization, 178
  organization, 158
  scenario, 161–162
  SMC prediction (*see* Smart medical centres (SMCs))
  study area, 173
  workflow, 162–163
Smart medical centres (SMCs), 157, 160
  crossover and mutation, 170–171
  establishment of, 174
  genetic algorithm, 164–165, 174
  multi-criteria objective model for, 171–172
  objective functions, 167–170
  population initialization, 165
  selection, 165–167
Soil moisture sensor, 108
Statistical analysis
  2^4 facorial design, 33
  Analysis of variance (ANOVA), 32
  experimental matrix, 34
  Henry line plot, 32
  Lenth method, 32
  normal probability plot, 39
  piroxicam/artichoke waste system, 33, 35
  residual distribution graph, 40
Stoichiometric and kinetic parameters, 20
Support vector machine (SVM), 86, 131
Surface and river water quality monitoring systems, 7–10

T

Temperature sensors, 107
Thermal imaging
  unmanned aerial vehicle and, 146–147
    crop water stress indicator, measurement of, 148
    water stress cumulative integration, 147
Thermal infrared sensors, 144–145
Thermoelectricity
  Boltztrap package, 45
  computation method, 45–46
  dicalcium germanide, 44
  dicalcium silicide, 44
  electrical conductivity, 47
  electronic thermal conductivity, 47–48
  energy band structure, 46–47
  figure of merit (ZT), 44

materials, 44
power factor, 49–50
Seebeck coefficient, 48–49
Three-dimensional response surface, 62, 64–66
Topology, 3
Tournament selection method, 167
Tran-Blaha-modified Becke-Johnson (TB-mBJ) exchange potential, 45
Transmission mode, 3
Transportation cost, 169
Transvaginal ultrasound (TVS) assessment, 90
Travelling salesman problem (TSP), 164

U

UAV anticipate, 140
UAV control system (UAV CS), 142
Unmanned aerial vehicle
precision agriculture using, 143
data acquisition, 143–145
machine learning techniques, 146
photogrammetry techniques, 145
UAS data processing, 145
vegetation indices calculation, 145
remote sensing using, 140–141
advantages, 142–143
data collection using sensors, 142
ground control station, 142
surveillance of crops, 141
UAV control system, 142
and thermal imaging, 146–147
crop water stress indicator, measurement of, 148
water stress cumulative integration, 147
Urban wastewater decontamination, 19
US Environmental Protection Agency (EPA), 118

V

Vegetation Index (VIs) calculation, 145
Visible light sensors (RGB), 144

W

Wastewater treatment plants (WWTP)
activated sludge process, 20, 24
ASM1 model, 21–22
biological and hydraulic simulations, 22
BOD_5 concentrations, 24
calibration of a model, 23, 24
COD concentrations, 24, 25
drug removal, 29
GPS-X software, 21, 22
input effluent characteristics, 22–23
stoichiometric and kinetic parameters, 20
Water contamination, 1
Water flow sensor, 108
Water quality
data processing, 3–4
features of, 4
models, 118
monitoring systems
aquaculture, 12–14
drinking water distribution networks, 10–12
natural water resources, 5–7
surface and river water, 7–10
traditional methods, 2
networking technologies, 3
water parameters
conductivity sensor, 3
dissolved oxygen sensor, 2–3
pH sensor, 2
temperature sensor, 3
turbidity sensor, 2
Water stress cumulative integration, 147
Watercourses, 20
Wearable healthcare sensors, 87
Western Rif of Morocco, limestone chain of, 117–119
coefficients of biochemistry action, 121–122
concentrations, simulation and monitoring of, 122–123
materials and methods, 119–120
simulation, 125
stability analysis, 123–125
study area, 119
water quality status, 120–121
Within set sum of squared errors (WSSSE), 96
WQI-NSF, 124, 125

Z

Zigbee modules, 9, 12, 105, 160, 161